Multiple Choice Questions in Biochemistry

Multiple Choice Questions in Biochemistry:

with answers and explanatory comments

D. G. O'Sullivan
W. R. D. Smith

Courtauld Institute of Biochemistry,
The Middlesex Hospital Medical School,
London

Edward Arnold

Introduction

Since the introduction of multiple choice question (MCQ) papers into Britain in the 1950–1960 decade, MCQs have gained general acceptance and today there must be few courses in any subject at any level that do not incorporate MCQs at some stage. They enable each student to be tested rapidly in an extensive range of topics and enable large numbers of students to be tested in an economical manner. They are also important as educational aids, as they provide a method for informing both students and tutors of inadequacies in the attention that has been paid to particular areas of the subject. This booklet has been designed, not only to provide students with practice in responding to MCQ-type questions, but to highlight, and help remedy, weaknesses. The explanations and hints accompanying the answers should help in these respects.

Two fundamental difficulties beset the constructor of MCQs. One is the problem of avoiding ambiguity in the questions and the second arises from the fact that many simply posed questions cannot be answered unequivocally with a **true** or **false** response. Such questions would have to be hedged with extensive qualifications and restrictions that would be inappropriate to the accepted style of MCQs. Naturally we have had these points very much in mind during the preparation of this booklet.

The general scheme adopted is that used in the companion booklet *Multiple Choice Questions in Physiology* by Bindman, Jewell and Smaje. A (usually brief) introduction or stem to each question is followed by 5 options or sub-questions, any or all of which could be true. A response **true** or **false** should be made for every option. Within this framework, however, some minor variation in style has been introduced deliberately. Thus, sometimes the statements are made without a specific question being asked, the query **true** or **false** being implied. Questions also vary in character, some being readily answerable if the appropriate background work has been assimilated and some requiring much greater thought and application of biochemical principles with, perhaps, a calculation. The questions are, as far as possible, arranged in topics, not in order of difficulty. Each question must be taken as standing on its own, in that answers occasionally contain hints to responses in later questions. The student is expected to cover over the answer page and provide responses for a complete question, and preferably responses for all the questions in a *section*, before referring to the right hand pages. The methods used by examiners for marking MCQ examinations vary and many are computer marked. Whilst marks are gained by correct responses, they are always lost by incorrect responses, but not always in a 1 to 1 ratio. As with all types of examination, practice at answering MCQ's will help students acquire confidence and skill.

We thank Professor P. N. Campbell, Director of the Courtauld Institute of Biochemistry, for his interest in this project.

D. G. O'Sullivan
W. R. D. Smith

Contents

Abbreviations

Units

g	gram
l	litre
m	metre
M	moles per litre
Pa	Pascal (1 Pa $\equiv$ 1 newton/m^2)
K_a	Ionization constant of an acid
K_m	Michaelis constant of an enzyme

Prefixes

k	kilo
m	milli
μ	micro
n	nano

Note
Substituted ammonium ions are symbolized as $R\overset{+}{N}H_3$. This enables the formula to be written backwards ($H_3\overset{+}{N}R$), as would frequently be necessary with amino acids and polypeptides, without introducing problems with the siting of the positive charge.

The ammonium ion itself is given the conventional NH_4^+ formula.

With reference to Questions 87 and 89, some test strips (such as Uristix) combine more than one test.

1 The structure of aspirin is

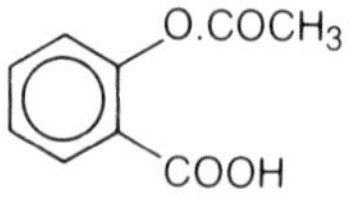

C_6H_4 (1-$O.COCH_3$, 2-$COOH$)

Which of the following statements are true for this compound and which are false?
(a) It is an aromatic carboxylic acid.
(b) It is a phenol.
(c) It is the acetyl derivative of a phenol.
(d) It is a 1,2-disubstituted benzene.
(e) It is called salicylic acid.

2 Consider the following structure and then note which of the statements are true and which are false.

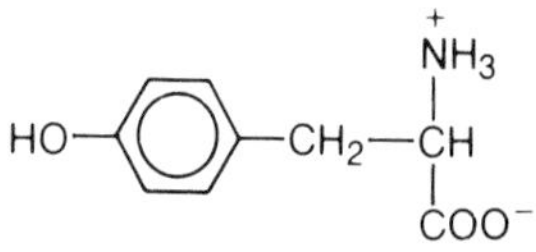

$HO-C_6H_4-CH_2-CH(\overset{+}{N}H_3)-COO^-$

(a) It is an aromatic amino acid.
(b) It is a phenol.
(c) It contains 2 asymmetric carbon atoms.
(d) It is a derivative of phenylalanine.
(e) The structure represents the compound tryptophan.

3 Consider glutathione, which possesses the following conventional structure:

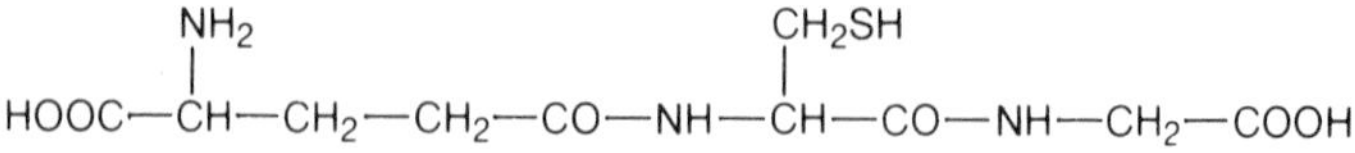

$$HOOC-CH(NH_2)-CH_2-CH_2-CO-NH-CH(CH_2SH)-CO-NH-CH_2-COOH$$

The following are statements about the molecule.
(a) An α-amino group is present.
(b) A link exists between 2 amino acid residues that differs from the usual peptide link.
(c) A readily oxidizable side-chain group is present.
(d) Two secondary amine groups are present.
(e) Glu-Cys-Gly would properly represent the structure.

1 (a) **True** Note the symbol for a benzene ring.
(b) **False** No OH group is attached to the benzene ring.
(c) **True** The phenol would be *o*-hydroxybenzoic acid (salicylic acid).
(d) **True**
(e) **False** Aspirin is acetylsalicylic acid.

2 (a) **True** It has an aromatic ring and contains both amino and carboxyl groups.
(b) **True** The OH group is attached directly to the aromatic ring.
(c) **False** It contains only 1 asymmetric carbon atom.
(d) **True** It is the *p*-hydroxy derivative of phenylalanine.
(e) **False** Tryptophan has an indole ring. This compound is the amino acid tyrosine.

3 (a) **True**
(b) **True** The glutamyl-cysteine linkage is *via* the γ-carboxyl and not the α-carboxyl group of the glutamine.
(c) **True** The $-CH_2SH$ group is readily oxidized to $-CH_2SO_3H$.
(d) **False** The NH groups in the chain are both linked to CO groups and are thus *amide* type groups rather than *amine* type groups.
(e) **False** Refer to the comment under (b) above.

4 Consider the two structures:

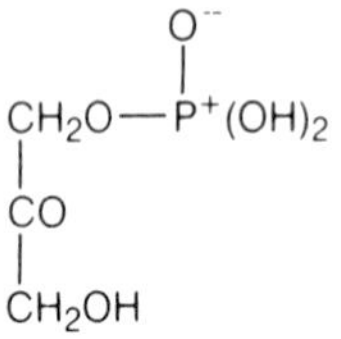

(A): $CH_2O{-}P^+(OH)_2$ (with O^- on P) | CO | CH_2OH

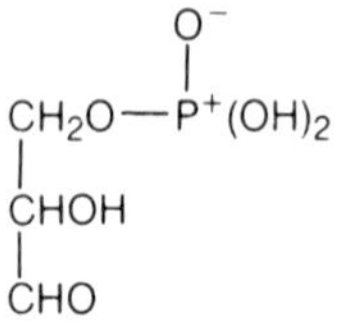

(B): $CH_2O{-}P^+(OH)_2$ (with O^- on P) | CHOH | CHO

(a) Compound (A) is a primary alcohol and compound (B) is a secondary alcohol.
(b) Compound (A) is a ketone and compound (B) is an aldehyde.
(c) (A) and (B) are enantiomers.
(d) (A) and (B) are not enantiomers, but are both optically active.
(e) (A) and (B) are triose phosphates.

5 The α-forms of D-glucose and D-galactose are hexoses that are closely related structurally. These two are called epimers (or an epimeric pair). Which of the following statements apply to these structures?
(a) They are mirror images of each other.
(b) They differ in configuration at one carbon atom only.
(c) One is an aldose and the other a ketose.
(d) They are structural isomers.
(e) They must both rotate the plane of polarization of plane-polarized light in a clockwise (+) direction.

6 Which statements relate to the following structure?

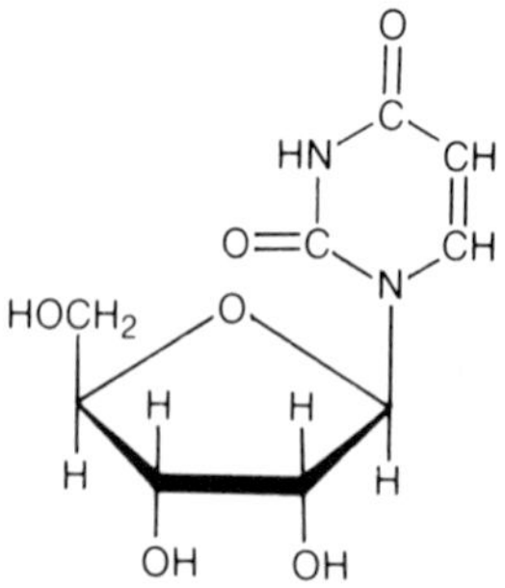

O, C, HN, CH, CH, O=C, N, HOCH2, O, H, H, H, H, OH, OH

(a) The compound is uridine.
(b) The compound is uracil.
(c) The structure is a nucleoside.
(d) The structure is a nucleotide.
(e) The pentose sugar is deoxyribose.

4 (a) **True** (A) contains the group $-CH_2OH$ and (B) contains the group $>CHOH$.
(b) **True** (A) contains the group $>CO$ and (B) contains the group $-CHO$.
(c) **False** They are isomers, but not optical isomers.
(d) **False** Only (B) has an asymmetric carbon atom. No optical activity is possible with (A).
(e) **True** (A) is the simplest ketose phosphate and (B) the simplest aldose phosphate.

5 (a) **False** The mirror image of D-glucose is L-glucose and of D-galactose is L-galactose.
(b) **True** The difference is at carbon atom 4 only.

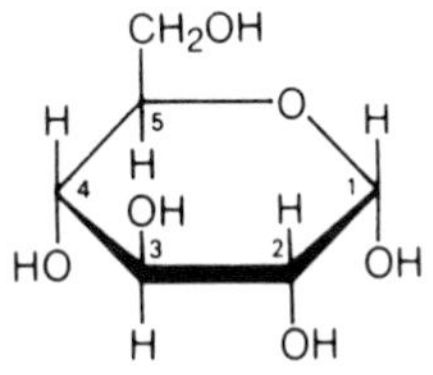

α-D-glucose

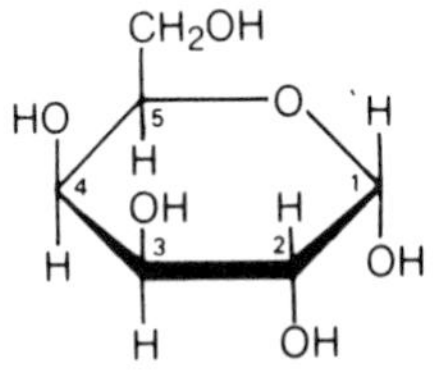

α-D-galactose

(c) **False** Both are aldoses.
(d) **True**
(e) **False** A D configuration compound does not necessarily rotate the plane of polarization in a clockwise direction.

6 (a) **True**
(b) **False**
(c) **True** Uridine is a nucleoside (in this case a ribosyl derivative of uracil).
(d) **False** A nucleotide is a nucleoside phosphate.
(e) **False** The pentose is ribose not deoxyribose.

7 Which statements are true and which are false concerning the structure:

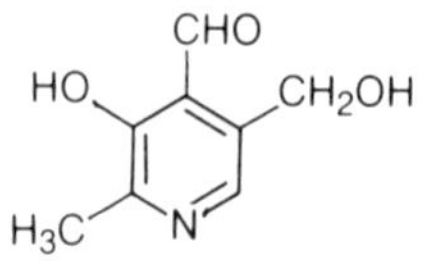

(a) It is a pyrimidine derivative.
(b) It is not readily oxidized.
(c) It possesses some aromatic character.
(d) It contains a formyl group.
(e) It is a base.

8 Cholesterol has the following structure:

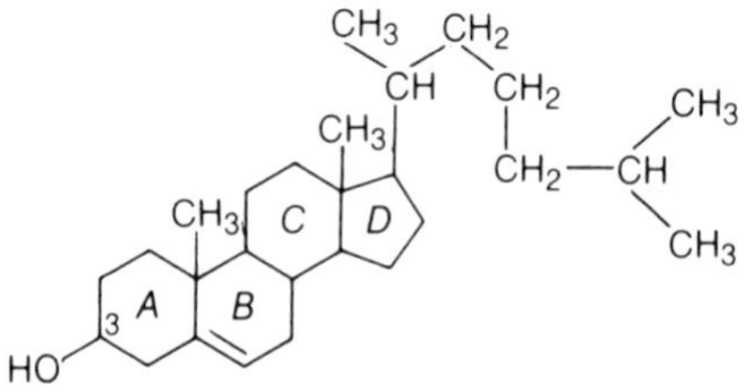

(a) It is an alcohol.
(b) It is a phenol.
(c) It contains an isopropyl group.
(d) It is a fatty acid.
(e) It is an unsaturated compound.

9 Linoleic acid has the structure (A)

$$CH_3{-}(CH_2)_4{-}CH{=}CH{-}CH_2{-}CH{=}CH{-}(CH_2)_7{-}COOH \qquad \text{(A)}$$

in which the brackets signify straight chains of 4 and of 7 CH_2 groups respectively.
(a) Linoleic acid has double bonds at the $\Delta 9$ and $\Delta 12$ positions.
(b) Linoleic acid has double bonds at the $\omega 6$ and $\omega 9$ positions.
(c) Structure (A) defines a particular molecule uniquely (i.e., without ambiguity).
(d) Linoleic acid could be called a polyunsaturated fatty acid.
(e) Linoleic acid might be obtained by reduction of stearic acid $C_{17}H_{35}.COOH$.

7 (a) **False** It is pyridoxal, which is a pyridine derivative.
(b) **False** It is readily oxidized, both enzymically and chemically, to 4-pyridoxic acid.
(c) **True** The pyridine ring is aromatic in character.
(d) **True** It is an aldehyde and thus contains the formyl group −CHO.
(e) **True** The nitrogen atom of the pyridine ring can accept a proton as shown:

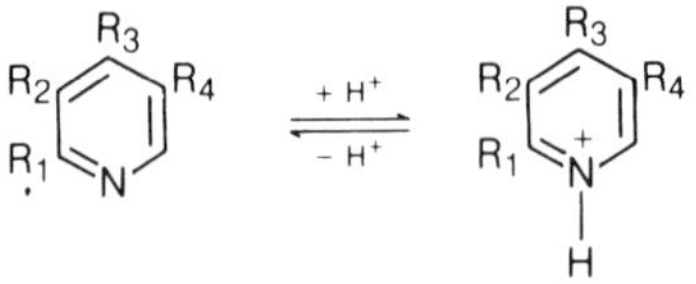

8 (a) **True** It is a secondary alcohol because of the −OH group attached to the alicyclic ring *A*. The group at position 3 is thus the >CHOH group.
(b) **False** The only −OH group present is that at position 3. This is not attached to an aromatic ring.
(c) **True** The group at the end of the long side-chain attached to ring *D* is an isopropyl group.
(d) **False** It is a lipophilic molecule, but *not* a fatty acid.
(e) **True** Ring *B* contains a double bond.

9 (a) **True** Numbering of the C-atoms starts at the carboxyl group.
(b) **True** Numbering of the C-atoms starts at the terminal methyl group.
(c) **False** *Cis–trans* isomerism is possible at each double bond and thus the structural formula (A) corresponds to four different stereoisomers: *cis–cis, cis–trans, trans–cis* and *trans–trans*.
(d) **True**
(e) **False** Stearic acid is a reduction product of linoleic acid not *vice versa*.

10 Consider the structure of protoporphyrin (haem without its iron atom):

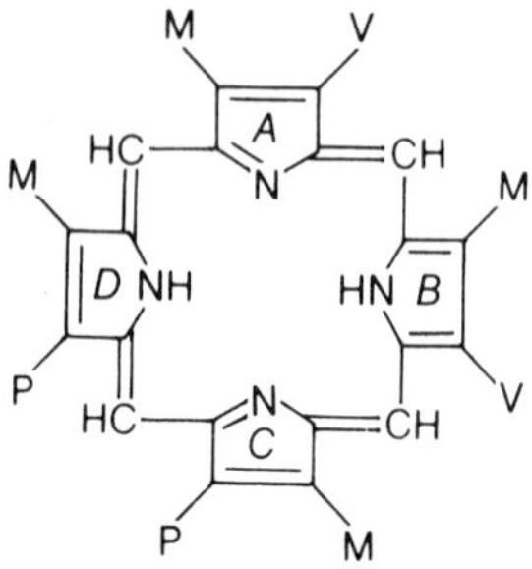

M = $-CH_3$
V = $-CH{=}CH_2$
P = $-CH_2.CH_2.COOH$

The colour of protoporphyrin is primarily related to
(a) The presence of the four pyrrole rings *A*, *B*, *C* and *D*.
(b) The particular pattern of the substituents M, V and P.
(c) The presence of the four bridging CH groups.
(d) The continuous conjugated double-bond system present in the structure.
(e) The cross-like configuration of the four pyrrole rings.

11 Changing the pH of an aqueous solution of histidine from pH 6 to its isoelectric point gives histidine molecules with no net charge. In some texts these uncharged molecules are given formula (A) and in others formula (B).

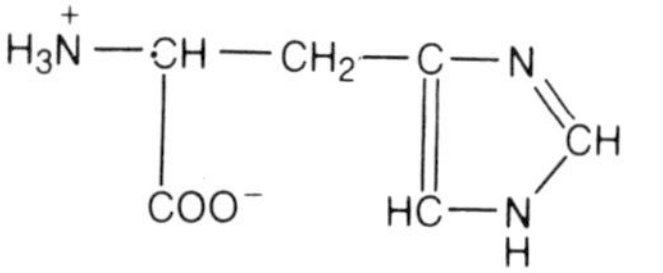

(A)

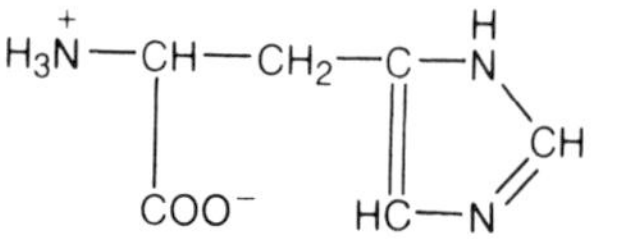

(B)

(a) Only one of these formulae could possibly be correct.
(b) Histidine contains only two asymmetric carbon atoms.
(c) The predominant forms of histidine present at the isoelectric point can act as bases.
(d) Histidine is an example of a heterocyclic compound.
(e) The protonated structures, formed by the addition of H^+ to structures (A) and (B), are sometimes called mesomeric forms (or mesomers).

10 (a) **False** The four rings *A, B, C* and *D* will not give rise to appropriately long wavelength absorption of light unless there is conjugation present.

(b) **False** This has little relevance to the colour.

(c) **False** Again, this will not give rise to colour without extensive conjugation.

(d) **True** The continuous conjugated double-bond system (the alternate single- and double-bond arrangement that is present) is the key feature in the system concerned with its colour. If other features are retained, but the conjugation broken, then the deep porphyrin type colour is removed.

(e) **False** This will impart no colour properties in the absence of a conjugated double-bond system.

11 (a) **False** Even at this low H^+ molarity (at pH 7·6), protons are continually being added to, and then dissociating from, structures (A) and (B). This occurs at a rapid rate. Addition of a proton to (A) and (B) give respectively

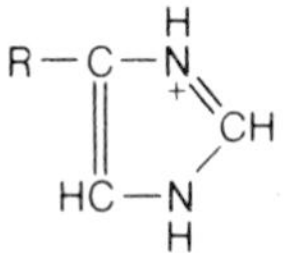

(C) 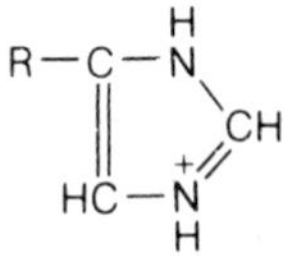 (D)

[R being the side-chain in (A) and (B)]. However, the actual ion will be a single entity not represented accurately by (C) and (D), but by an intermediate entity with the +ve charge shared between both N atoms in the ring. Structures (A) and (B) are known as tautomers (or tautomeric forms), both being present in approximately equal concentrations at the isoelectric point. Structures (C) and (D) are mesomeric forms representing a single entity.

(b) **False** It contains only one.

(c) **True** At the isoelectric point, the predominant structures will be (A) and (B). These will react with added hydrogen ions as indicated in (a) above.

(d) **True** The 5-membered ring (called an imidazole ring) is a heterocyclic ring.

(e) **True** These are the structures (C) and (D). The relative positions of the atoms in (C) and (D) are almost identical and there is little difference in energy between (C) and (D). The difference between them is a matter of electron distribution. In these circumstances, neither (C) nor (D) accurately represents the actual molecular structure, which is intermediate between (C) and (D) and of lower energy than either. Structures (C) and (D) are sometimes called mesomeric forms (or mesomers).

12 Consider the molecular entities:

$CH_3.COOH$ $\quad$ $H_3\overset{+}{N}.CH_2.COOH$ $\quad$ $H_2PO_4^-$

$CH_3.CHOH.COOH$ $\quad$ H_2CO_3

Which of the following statements are correct?
(a) Only $CH_3.COOH$, $CH_3.CHOH.COOH$, and H_2CO_3 are acids.
(b) They are all acids except $H_2PO_4^-$.
(c) They are all acids.
(d) The strongest acid in the 5 entities is $CH_3.COOH$.
(e) $H_2PO_4^-$ is a base.

13 Consider the molecular entities:

(A) $H_3\overset{+}{N}.CH(COOH).CH_2.CH_2.CH_2.CH_2.\overset{+}{N}H_3$ $\quad$ (B) PO_4^{3-}

(C) OH^- $\quad$ (D) NH_3 $\quad$ (E) CN^-

Which of the following statements are correct?
(a) (C), (D), and (E) are bases, but not (A) and (B).
(b) All are bases except (A).
(c) All are bases.
(d) OH^- is the strongest base present in (A) to (E).
(e) NH_3 cannot be a base as it is uncharged.

12 (a) **False** This is in spite of the fact that only these 3 are commonly named in the style: acetic acid, lactic acid, carbonic acid.

(b) **False** $H_2PO_4^-$ is also an acid. It can donate a proton as illustrated by

$$H_2PO_4^- \rightleftharpoons H^+ + HPO_4^{2-}$$

Acid — Conjugate base

(c) **True**

(d) **False** The strongest acid is $H_3\overset{+}{N}.CH_2.COOH$. The proton-donating process concerned is

$$H_3\overset{+}{N}.CH_2.COOH \rightleftharpoons H_3\overset{+}{N}.CH_2.COO^- + H^+$$

Acid — Conjugate base

(e) **True** $H_2PO_4^-$ is not only an acid, as described above under (b), but it is also a base:

$$H_3PO_4 \rightleftharpoons H^+ + H_2PO_4^-$$

Acid — Conjugate base

13 (a) **False** A base is a proton acceptor, and so PO_4^{3-} is a base as well as OH^-, NH_3, and CN^-.

$$H^+ + PO_4^{3-} \rightleftharpoons HPO_4^{2-}$$

(b) **True** (A) is the fully protonated form of the basic amino acid lysine. Being fully protonated it cannot accept another proton. In fact (A) is an acid with pK_a of 2.18. All the other molecular species (B) to (E) are bases.

(c) **False**

(d) **True** The order of basic strength is

$$OH^- > PO_4^{3-} > NH_3 > CN^-$$

The pK_a of the reaction $H_2O \rightleftharpoons H^+ + OH^-$ is 15.74 (at 20°C). This very high value shows that OH^- is a very strong base.

(e) **False**

14 The graph shows the change in pH of 0.1 M orthophosphoric acid when 10 ml of the acid is titrated with sodium hydroxide solution.

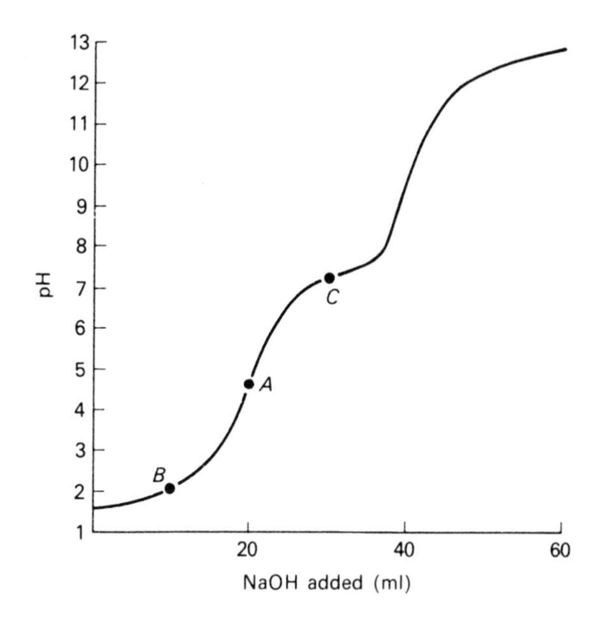

(a) Point *A* represents a solution which is 0.1 M monosodium orthophosphate.
(b) The logarithm of the reciprocal of the ionization constant for the first stage of the ionization of orthophosphoric acid pK_1 has the value 2.1.
(c) The pK_2 for orthophosphoric acid has the value 7.2.
(d) The most effective buffer solutions (between pH 3 and pH 11) achieved during the titration are at, and in the neighbourhood of, point *C*.
(e) The titrating sodium hydroxide solution is 0.025 M.

15 Take the pK_a of H_2CO_3 to be 6.1 and of $H_2PO_4^-$ to be 7.2. Which of the following statements are true and which are false for aqueous solutions containing only those dissolved components stated?
(a) A molar solution (1.0 M) of $NaHCO_3$ has a pH greater than 7.
(b) A molar solution of Na_2HPO_4 has a pH greater than 7.
(c) A molar solution of NaH_2PO_4 has a pH greater than 7.
(d) A solution which is equimolar with respect to H_2CO_3 and $NaHCO_3$ has a pH less than 7.
(e) The presence of H_2CO_3 and HCO_3^- and also $H_2PO_4^-$ and HPO_4^{2-} adequately control blood pH at 7.41, no other buffer system being necessary.

14 (a) **False** Point *A* represents approximately 0.033 M monosodium orthophosphate.

(b) **True** This is the pH of a solution that is equimolar with respect to the acid and its monosodium salt, i.e. the pH of point *B*.

(c) **True** This is the pH of a solution that is equimolar with respect to the mono- and disodium salts, i.e. the pH of point *C*.

(d) **True** Point *C* is the monosodium orthophosphate – disodium orthophosphate buffer possessing equal concentrations of the ions $H_2PO_4^-$ and HPO_4^{2-}.

(e) **False** The sodium hydroxide is 0.05 M.

15 (a) **True** As we take the pK_a of H_2CO_3 as 6.1, it follows from the Henderson–Hasselbalch equation that the pH of a solution containing 1.0 M $NaHCO_3$ and 0.1 M H_2CO_3 must be 7.1. Clearly, the pH of 1.0 M $NaHCO_3$ must be greater than 7.1. An alternative approach is as follows: according to the relationship for the Na^+ salt of a weak acid HA:

$$pH = \tfrac{1}{2}pK_w + \tfrac{1}{2}pK_a + \tfrac{1}{2}\log_{10}[\text{base}],$$

the pH of 1.0 M $NaHCO_3$ should be approximately 10. This assumes the absence of dissolved CO_2 other than that arising from the $NaHCO_3$.

(b) **True** Again the Na^+ salt of a weak acid will be alkaline due to reaction with water. The affinity for H^+ ions of the HPO_4^{2-} ions results in an increase in $[OH^-]$ in the solution.

(c) **False** The $H_2PO_4^-$ ion has little affinity for protons and it acts as a proton donor.

(d) **True** The Henderson–Hasselbalch equation

$$pH = pK_a + \log([HCO_3^-]/[H_2CO_3])$$

gives a pH of 6.1.

(e) **False** The buffering action of protein in blood is essential.

16 Which of the following statements are true and which false?
(a) The pH of a buffer solution remains constant if any quantity of an acid or base is added.
(b) The carbonic acid–bicarbonate system is the sole chemical process that acts to transport CO_2 from the tissues to the lungs.
(c) Increasing the CO_2 content of an aqueous solution raises its pH.
(d) Changing the pH cannot alter the properties of the active site of an enzyme.
(e) Ketosis, whether due to diabetes, starvation or a high-fat diet tends to lead to a metabolic acidosis.

17 Which of the following statements are correct and which false in relation to the *uncompensated* states of acidosis and alkalosis in human plasma shown in the graph? Region *N* represents a normal acid–base status. The plasma was obtained from arterial blood anaerobically.

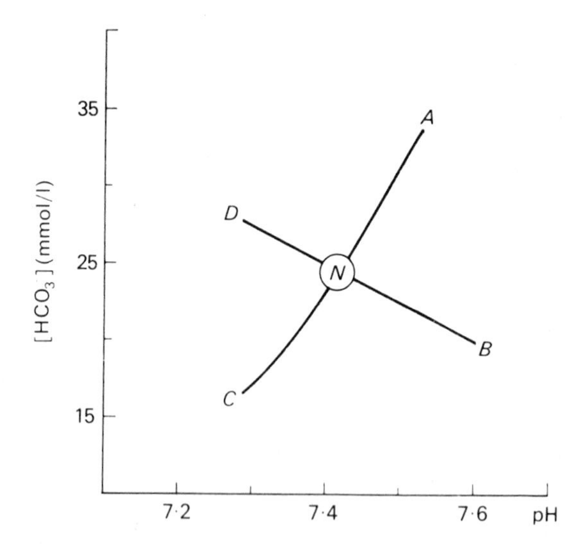

(a) Path *NA* represents the development of respiratory alkalosis.
(b) Path *NB* represents the development of metabolic alkalosis.
(c) Path *NC* represents the development of metabolic acidosis.
(d) Path *ND* represents the development of respiratory acidosis.
(e) Line *ANC* must be within the normal range of the 5.3 kPa (40 mm Hg) Pco_2 isobar line.

16 (a) **False** Continued addition of acid or base produces a small but cumulative change in pH of a buffer solution until the pH reaches a figure of $pK_a \pm 1$ (pK_a referring to the buffer acid component). Further addition of acid or base increases the rate of change of pH markedly.

(b) **False** Human haemoglobin, particularly the deoxy form, at pH 7.4 has uncharged α-amino groups that can take up and subsequently relinquish CO_2 molecules:

$$R{-}NH_2 + CO_2 \rightleftharpoons R{-}NH.COO^- + H^+$$

(c) **False** It will tend to lower the pH.

(d) **False** Any ionized or ionizable groups at the active sites will be influenced by change in pH. Changing the ionization status at an active site is bound to alter the properties of the site.

(e) **True** Increased plasma levels of acetoacetic and hydroxybutyric acids will produce metabolic acidosis, but compensation may reduce the degree of acidosis substantially.

17 (a) **False** Path *NA* represents metabolic alkalosis. This is a state with increased plasma pH and increased plasma $[HCO_3^-]$ and which must follow a path in which P_{CO_2} is constant (and normal) if no respiratory component is involved.

(b) **False** Path *NB* represents respiratory alkalosis. This is a state with increased plasma pH and reduced $[HCO_3^-]$. It must follow the normal buffer line and thus have no base excess or deficit outside a normal range.

(c) **True** Entry into the blood of abnormally large amounts of acids that are stronger acids than H_2CO_3, not only lower the pH, but react with HCO_3^- ions and reduce $[HCO_3^-]$.

(d) **True** Increase in CO_2 and H_2CO_3 concentrations, because of ionization of the H_2CO_3, results in reduction in pH and increase in $[HCO_3^-]$.

(e) **True**

18 The graph shows pathways of deviation from normal conditions relating plasma base excess or deficit (milliequivalents per litre) with its CO_2 pressure (mm Hg). The plasma was obtained from arterial blood anaerobically.

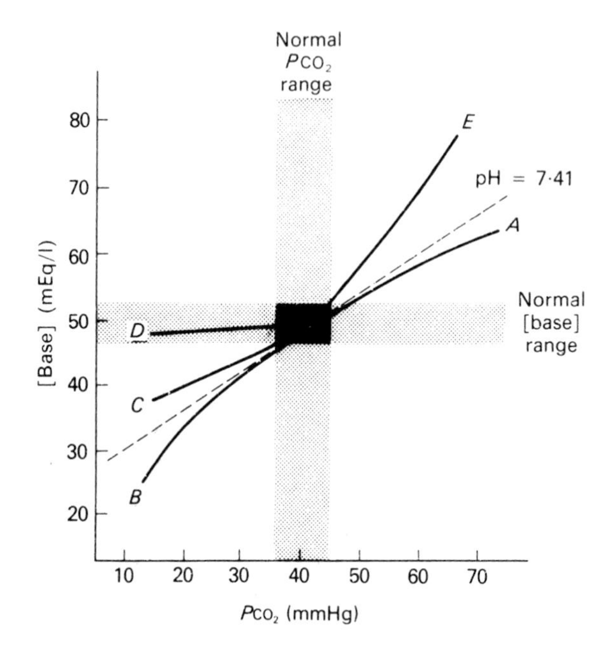

(a) Path *A* represents respiratory acidosis with renal compensation.
(b) Path *B* represents metabolic alkalosis with respiratory compensation.
(c) Path *C* represents respiratory alkalosis with renal compensation.
(d) Path *D* represents respiratory alkalosis without significant renal compensation.
(e) Path *E* represents metabolic acidosis with respiratory compensation.

18 (a) **True** Areas below the broken line (pH = 7.4) represent acidosis. Uncompensated respiratory acidosis would lie on or close to the CO_2 excess axis, whilst compensation would divert the path nearer to the broken line.

(b) **False** This pathway represents the development of a metabolic acidosis with respiratory compensation.

(c) **True** Respiratory alkalosis will be represented by a line near the CO_2 deficit axis. Renal compensation will divert this towards the broken line.

(d) **True**

(e) **False** Path *E* represents metabolic alkalosis, with respiratory compensation diverting the uncompensated line nearer to the broken line.

19 The graph shows the changes in pH that occur during the titration of a solution of 1.0 millimole of glutamic acid with HCl and with NaOH solutions.

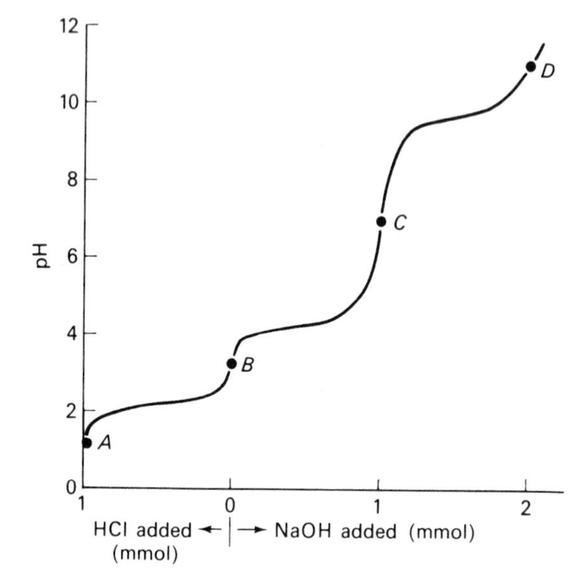

(a) At point *A*, the glutamic acid is mainly in structural form (P)

(P) $H_2N—CH(COO^-)—CH_2—CH_2—COO^-$

(b) At point *B*, the glutamic acid is mainly in structural form (Q)

(Q) $H_3\overset{+}{N}—CH(COO^-)—CH_2—CH_2—COO^-$

(c) At point *C*, the glutamic acid is mainly in structural form (R)

(R) $H_3\overset{+}{N}—CH(COO^-)—CH_2—CH_2—COOH$

(d) At point *D*, the glutamic acid is mainly in structural form (S)

(S) $H_3\overset{+}{N}—CH(COOH)—CH_2—CH_2—COOH$

(e) If the pHs at points *B* and *C* are 3.22 and 6.96 respectively and pK_2 (the pK_a for the intermediate of the 3 ionization stages) is 4.25, then $pK_1 = 2.19$ and $pK_3 = 9.67$.

19 (a) **False** At point *A* the glutamic acid will be 'fully' protonated, i.e. have predominantly structure (S) with only a minute proportion of other structures.

(b) **False** Point *B* is the end of the first ionization stage. At this point the predominant structure will be (R).

(c) **False** Point *C* is the end of the second ionization stage and the predominant structure will be (Q).

(d) **False** Point *D* is the end of the third ionization stage with the predominant structure (P)

(e) **True** A pK_a is the pH at the midpoint of an ionization stage. If more than 1 stage exists, it can be shown that the equivalence pH is the mean of the pK_a values for the stages before and after this equivalence point. Thus

$$pK_1 = 3.22 - (4.25 - 3.22) = 2.19$$

$$pK_2 = 6.96 + (6.96 - 4.25) = 9.67.$$

20 An experimenter's incomplete notes inform us that a set of eight colourless protein solutions covering a range of concentrations were prepared and 1.0 ml of each was allowed to react with 4.0 ml of biuret reagent. A standard light absorbance against protein concentration curve was plotted. Study the curve and decide which of the following statements are true and which false?

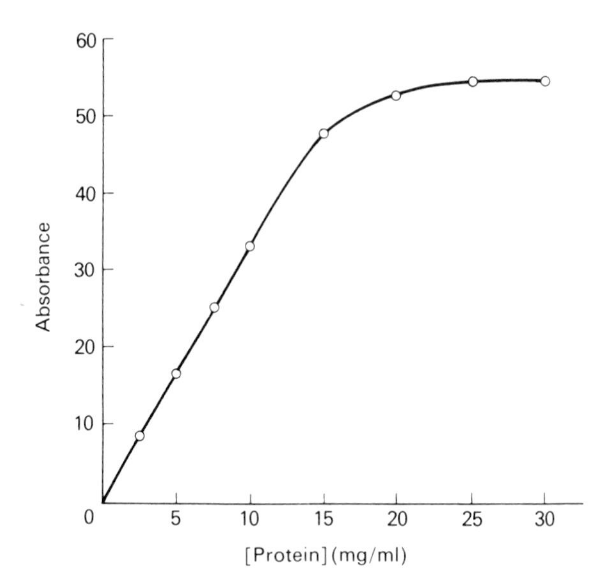

(a) The graph proves that no 'blank' solution had been prepared and used in the experiment.
(b) The graph proves that an unsuitable wavelength range filter had been used in the experiment.
(c) It is possible that adequate biuret reagent had not been supplied to all the samples.
(d) The Beer–Lambert Law was not applicable to the reaction in any protein concentration range.
(e) Absorbance of 20 was recorded for one of the experimenter's samples. This would imply a protein concentration of 6 mg/ml.

21 Which of the following techniques involve measurement of the emission of visible or ultraviolet light?
(a) Colorimetry.
(b) Phosphorimetry.
(c) Fluorescence spectrometry.
(d) Nuclear magnetic resonance spectrometry.
(e) Flame photometry.

20 (a) **False** The biuret reagent is coloured and it is very doubtful that measurements without employing a 'blank' would provide a line passing through the origin.
(b) **False** There is no evidence from the graph that an unsuitable optical filter was used.
(c) **True** This would certainly account for the failure of the graph to continue to show an increase in absorbance at higher protein concentrations.
(d) **False** The evidence suggests that the Beer–Lambert Law is obeyed well at the lower concentrations.
(e) **True** The graph shows that this is true well within the limits of accuracy implied by the absorbance and concentration data in (e).

21 (a) **False** Colorimetry involves measurement of light absorption.
(b) **True**
(c) **True**
(d) **False** Absorption of radiowave frequency radiation is measured, the sample concerned being subjected to a magnetic field.
(e) **True**

22 Two pure substances, X and Y, isolated from different sources, have the same R_F value in a chromatographic system. Which of the following are valid:
(a) X and Y are possibly identical.
(b) X and Y are certainly identical.
(c) X and Y are certainly different.
(d) X and Y are similarly ionized.
(e) X and Y have identical molecular weights.

23 Sephadex is a cross-linked dextran molecular matrix. Consider a column of Sephadex gel (exclusion limit = 5000 mol. wt.), total volume 100 ml, of which 60 ml is the internal volume of gel particles. The top of the column is loaded with a mixture of compounds X, Y, and Z in solution and eluted with solvent. X is eluted from the column when about 35 ml has passed through, Y when 100 ml and Z after 150 ml. Which of the following statements can be expected to apply?
(a) Z must be a carbohydrate.
(b) The molecular size of Y must be greater than that of X.
(c) The solubility in solvent of Z must be less than that of X and Y.
(d) Compound Z is adsorbed to the gel material.
(e) If X was albumin, Y could be vasopressin.

24 A protein was found to contain no tryptophan or histidine and no unusual amino acids. Enzyme hydrolysis gave a mixture of peptides, a small sample of which was placed at the origin (point *X*) and subjected to electrophoresis along the paper in direction 1 as shown in the diagram. The paper was dried and subjected briefly to an atmosphere containing the vapour of a strong oxidizing agent (e.g., performic acid). The oxidizing agent was permitted to evaporate away and the paper subjected to electrophoresis under the same conditions as before, but in direction 2 (at right angles to direction 1). Treatment with ninhydrin produced the pattern of spots shown in the diagram.

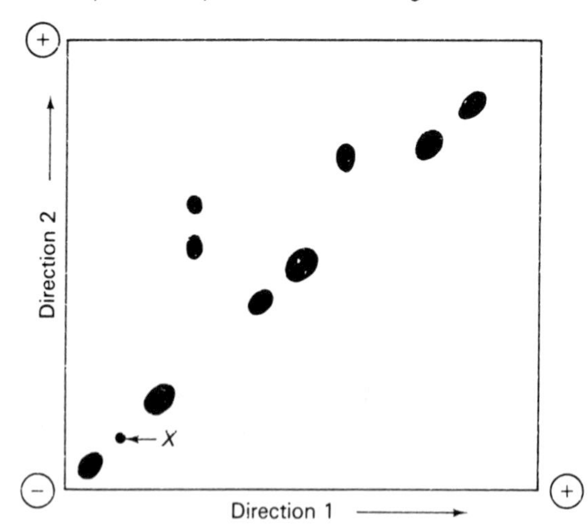

(a) The double spot, not on the diagonal, is produced by lysine and arginine.
(b) The first electrophoresis (direction 1) separated 8 peptides.
(c) No peptide has been changed by the oxidation procedure.
(d) One peptide possibly contains a disulphide bridge linking two dissimilar peptide chains.
(e) One peptide must contain a disulphide link formed from 2 cystine residues in the same peptide chain.

22 (a) **True** This would be the correct conclusion based on chromatography. Further investigation would be required.
(b) **False**
(c) **False**
(d) **False**
(e) **False**

23 (a) **False** The carbohydrate character of Sephadex does not imply that carbohydrates have to be strongly held in the column.
(b) **False** Separation in a Sephadex column depends mainly on molecular size, but large molecules are eluted first.
(c) **False** X, Y and Z are dissolved and their solubilities in the solvent are irrelevant.
(d) **True** The delay in elution of Z indicates some degree of adsorption.
(e) **True** Albumin is a protein of molecular weight above the exclusion limit for the sephadex. Vasopressin is a nonapeptide of molecular weight below the exclusion limit.

24 (a) **False** If two products of the enzymic hydrolysis happened to be single amino acids they would still necessarily lie on the diagonal line occupied by most peptides.
(b) **True** Two of these underwent modification during the subsequent oxidative procedure.
(c) **False**
(d) **True** Only a disulphide group would be capable of rapid oxidation to give ions with increased negative charge.

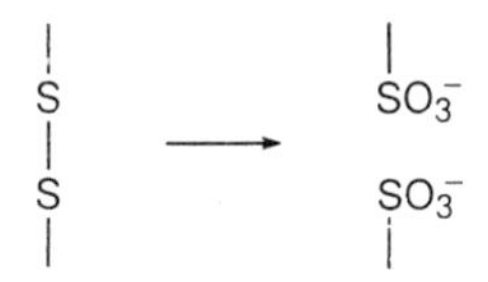

The 2 spots might be due to different peptides that before oxidation were joined by the –S–S– bridge. An alternative possibility is that 2 peptides, both with –S–S– bridges, failed to separate in the first electrophoresis and provided the 2 spots in the second.
(e) **False** The statement would be correct if the word 'must' were replaced by 'might'. Another possibility is that 2 separate, but identical, peptide chains might have been joined by an –S–S– bridge.

25 A large excess of cation-exchange resin was fully charged with K^+, containing traces of radioactive $^{42}K^+$ (possessing a half-life of 12.5 hours). The resin was washed, suspended in water, and then treated with 10 ml of 1.0 M calcium chloride. The displaced K^+ registered a radioactivity of 30 000 counts per minute in a counter with 50% efficiency.
(a) The released potassium totalled 0.42 g.
(b) The released potassium totalled 20 millimoles.
(c) The specific activity of the released potassium was 10 disintegrations per minute per micromole.
(d) The specific activity of the released potassium was 3×10^6 disintegrations per minute per mole.
(e) The liberated potassium only registered 7500 counts per minute, 50 hours later.

26 Which of the following statements are true and which are false for affinity chromatography?
(a) It is another name for gas-liquid chromatography.
(b) It is a name applied to techniques that combine partition chromatography and electrophoresis.
(c) It is a method that has only been applied to proteins.
(d) It has proved valuable for purifying enzymes.
(e) Its disadvantage is lack of specificity in separating macromolecules.

27 The following statements concern the use of isotopes.
(a) ^3H, ^{14}C and ^{32}P are all β-particle emitters.
(b) Deuterium is used less frequently in biochemistry than tritium because more radiation precautions are required in the use of deuterium.
(c) The production of photons by emitted particles is sometimes employed in determining the disintegrations per minute of radioactive material.
(d) The transference of label from a tritium labelled substance to other hydrogen compounds never occurs.
(e) γ-Radiations do not ionize the contents of cells, but damage tissue solely by virtue of a heating effect.

28 Which statements are true and which false concerning subcellular fractions of mammalian cells separated by fractional centrifugation?
(a) DNA will be present in the nuclear fraction.
(b) The soluble fraction will contain microscopically visible fat droplets.
(c) Monoamine oxidases will be present in the mitochondrial fraction.
(d) The lysosome fraction will possess marked transaminase activity.
(e) The mitochondrial fraction will contain an ATP generating system.

25 (a) **False**
(b) **True** The displaced K^+ must be the equivalent of 10 ml of 1.0 M Ca^{2+}. The equivalent of 10 millimoles of Ca^{2+} is 20 millimoles of K^+.
(c) **False** As 60 000 disintegrations per minute occur from 20 millimoles, it follows that the radioactivity is 3 disintegrations per minute per micromole.
(d) **True** 3 disintegrations per minute per micromole corresponds to 3×10^6 disintegrations per minute per mole.
(e) **False** An initial 30 000 counts per minute will have dropped to 1875 counts per minute, 50 hours later.

26 (a) **False**
(b) **False**
(c) **False** It has been applied mainly to proteins, but is certainly not restricted to proteins.
(d) **True** The specificity of the active site of enzymes, and possibly of allosteric sites also, makes the method particularly appropriate for enzymes. A suitable competitive inhibitor is chemically attached to the supporting matrix, columns of which can then be used to separate and purify the enzyme.
(e) **False** Specificity is very high.

27 (a) **True** The energy of the emitted β-particle is very low for ^3H and low for ^{14}C.
(b) **False** Deuterium (^2H) is not radioactive.
(c) **True** The solid scintillators used in nuclear medicine comprise single crystals of thallium-activated sodium iodide. Liquid scintillators are solutions of fluorescent oxazole derivatives in toluene or similar solvents.
(d) **False** How readily it occurs depends on how labile the chemical bonds are that link the tritium and hydrogen atoms to their respective molecules. In many cases exchange occurs readily.
(e) **False**

28 (a) **True** Note, however, that not all the DNA is necessarily in the nucleus.
(b) **False** These should have separated.
(c) **True**
(d) **False**
(e) **True** Mitochondria will possess an ATP generating system (the respiratory chain and oxidative phosphorylation) although in particular cases it may be inhibited or uncoupled.

29 For which of the amino acids (a) to (e) is the following statement true and for which false?

Each of the amino acids listed below is a component of polypeptide chains in proteins (conventional chemical formulae are given):

(a) $H_2N.CH_2.CH_2.CH_2.CH(NH_2).COOH$

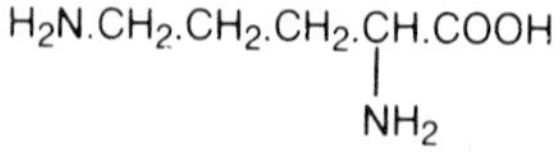

(b) $HS.CH_2.CH_2.CH(NH_2).COOH$

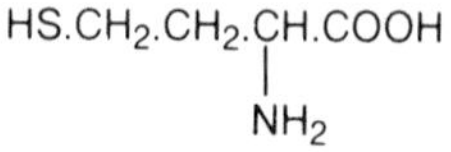

(c) Pyrrolidine ring: H_2C—CH_2 / H_2C and $CH.COOH$ joined through NH

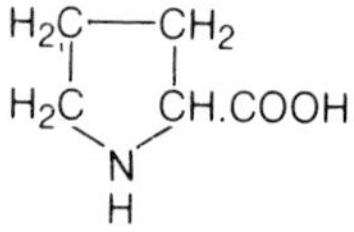

(d) $H_2N.CH_2.CH_2.COOH$

(e) $H_3C.CH(OH)—CH(NH_2)—COOH$

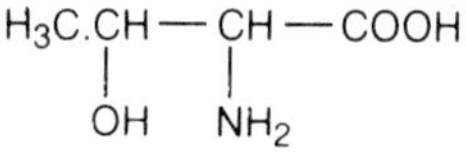

30 The structure

$$H_3\overset{+}{N}—C(H)(CH_2—C(=O)—NH_2)—COO^-$$

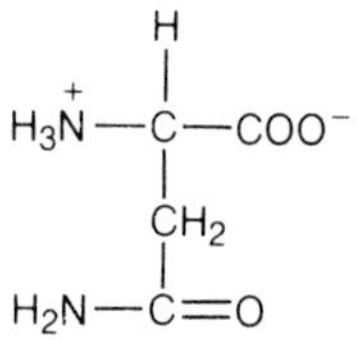

(a) Is a dipeptide.
(b) Is given the abbreviation GABA.
(c) Is given the abbreviation Asn.
(d) Can be involved in linking a polysaccharide chain to a protein chain.
(e) Is an essential dietary component for humans.

31 Which of the following α-amino acids have their corresponding α-oxo acids as intermediates of the citric acid cycle?
(a) Alanine.
(b) Aspartic acid.
(c) Valine.
(d) Glutamic acid.
(e) Ornithine.

29 (a) **False** This is ornithine, a component of the urea cycle.
(b) **False** This is homocysteine, the precursor of methionine in methionine biosynthesis.
(c) **True** This is proline.
(d) **False** This is β-alanine, it is present (as a residue) in part of the coenzyme A structure.
(e) **True** This is threonine.

30 (a) **False** It is an amino acid.
(b) **False** GABA is the abbreviation of gamma-aminobutyric acid, $H_2N.CH_2.CH_2.CH_2.COOH$.
(c) **True** Asn is the α-amino acid asparagine, the amide of aspartic acid.
(d) **True**
(e) **False** It can readily be synthesized from other amino acids.

31 (a) **False**
(b) **True** Oxaloacetic acid is the corresponding α-oxo acid.
(c) **False**
(d) **True** α-Oxoglutaric acid is the corresponding α-oxo acid.
(e) **False**

32 Which of the following are classified as 'essential amino acids' for man?
(a) Phenylalanine.
(b) Methionine.
(c) Glycine.
(d) Serine.
(e) Tryptophan.

33 Consider the compound represented by the sequence

Phe-Leu-Ala-Val-Phe-Leu-Lys

This compound will
(a) Have molecules possessing 7 peptide links.
(b) Be a basic peptide.
(c) Possess no affinity for lipid surfaces.
(d) Be hydrolysed by trypsin.
(e) Have an isoelectric point above pH 8.

34 The following processes are likely to be of value in determining the amino acid sequence of a peptide.
(a) Racemization.
(b) Reaction with 2,4-dinitrofluorobenzene.
(c) Hydrolysis with chymotrypsin.
(d) Reaction with ribonuclease.
(e) The Edman degradation.

35 Cleavage of disulphide bonds in proteins may be achieved by
(a) Dispersing the protein in 8 M urea.
(b) Dialysis.
(c) Reaction with performic acid (H.COOOH).
(d) Reaction with mercaptoethanol ($HS.CH_2CH_2OH$).
(e) Hydrolysis.

32 (a) **True**
(b) **True**
(c) **False** Glycine can be formed by metabolic processes in the body.
(d) **False** Serine can also be formed in the body.
(e) **True**

33 (a) **False** This is a heptapeptide with 6 peptide links.
(b) **True** Lysine is a basic amino acid whilst the others are 'neutral' amino acids.
(c) **False** As the only hydrophilic group in all 7 side-chains is the amino group at the end of the lysine side-chain and otherwise the side-chains are lipophilic, it follows that some measure of affinity for lipid surfaces is present.
(d) **False** Trypsin hydrolyses peptides on the carboxyl side of lysine and arginine residues. It will not hydrolyse this heptapeptide.
(e) **True** It is a basic peptide and the isoelectric point of lysine is considerably higher than 8 ($pI = 9.7$), consequently the isoelectric point of the peptide will be above pH 8.

34 (a) **False** Racemization is the conversion of a single optical isomer into a material which has 50% of each of the two enantiomers.
(b) **True** This can lead to the elucidation of the N-terminal amino acid.
(c) **True** This hydrolyses peptide bonds of specific amino acids.
(d) **False** Reaction should not occur.
(e) **True** Edman's reagent (phenylisothiocyanate) reacts with the N-terminal amino acid and a derivative of this acid can then be split off to leave the rest of the polypeptide chain intact.

35 (a) **False** This procedure will weaken or break hydrogen bonds, but will not affect disulphide bonds.
(b) **False** Dialysis will separate free or loosely bound small molecules, but will not break disulphide bonds in proteins.
(c) **True** The oxidizing action breaks the bonds, producing cysteic acid residues.
(d) **True** Here the action of the reagent is one of reduction.

$$\text{Chain}_1\text{—S—S—Chain}_2$$

$$\downarrow \text{HS—CH}_2\text{CH}_2\text{OH}$$

$$\text{Chain}_1\text{—S—S—CH}_2\text{CH}_2\text{OH} + \text{HS—Chain}_2$$

$$\downarrow \text{HS—CH}_2\text{CH}_2\text{OH}$$

$$\text{Chain}_1\text{—SH} + (\text{—S—CH}_2\text{CH}_2\text{OH})_2 + \text{HS—Chain}_2$$

(e) **False**

36 In each of five mutant haemoglobins, a single amino acid substitution in HbA has occurred as shown:

HbD	Asn	→	Lys	HbJ	Gly	→	Asp
HbN	Lys	→	Glu	HbC	Glu	→	Lys
Hb Sydney	Val	→	Ala				

Their relative electrophoretic mobilities are as follows:

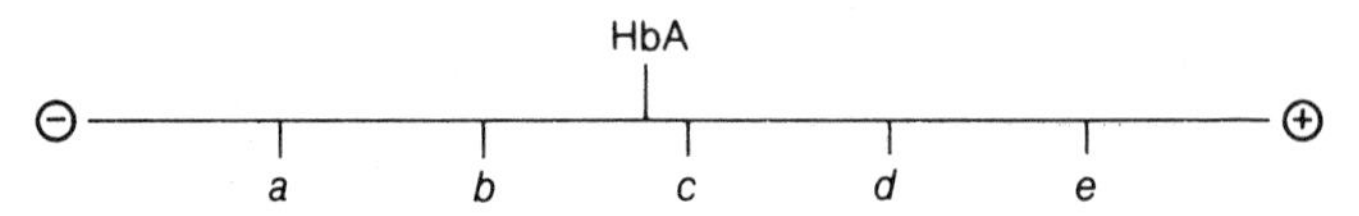

Which of the following statements are true and which false?
(a) Band *a* represents HbD.
(b) Band *b* represents HbC.
(c) Band *c* represents Hb Sydney.
(d) Band *d* represents HbJ.
(e) Band *e* represents HbN.

37 Which of the following statements are true and which false in relation to myoglobin?
(a) It is a compact globular structure consisting of a single polypeptide chain to which a single haem group is attached.
(b) The polypeptide chain has a high proportion of its length in an α-helical confirmation.
(c) Most non-polar residues lie on the exterior of the globular structure.
(d) The haem group is located between 2 histidine residues.
(e) If the haem group is removed from the apoprotein and the apoprotein unfolded without interfering with its primary structure, then there is no way of reversing this process.

38 Abnormal haemoglobins *may* differ from normal adult HbA in
(a) The presence of a D-amino acid instead of an L-amino acid.
(b) The presence of 4 identical chains.
(c) Possession of a different electrophoretic mobility.
(d) Replacement of a single amino acid residue by a different amino acid residue in one of the chains.
(e) Replacement of lysine by hydroxylysine.

36 (a) **False** HbD has one basic amino acid replacing a neutral amino acid.
(b) **False** HbC has one basic amino acid replacing an acidic amino acid. Considering (a) and (b) together it is clear that their order is incorrect so that both (a) and (b) must be false.
(c) **True** A neutral amino acid replacing another neutral amino acid.
(d) **True** An acidic amino acid replacing a neutral amino acid.
(e) **True** An acidic amino acid replacing a basic amino acid.

37 (a) **True**
(b) **True** Over 70% of the polypeptide is in the α-helical conformation.
(c) **False** The interior consists mainly of non-polar residues (with 2 histidines). Polar residues mainly lie in the exterior region.
(d) **True** The haem iron is covalently attached to only one, His F8. An O_2 molecule can attach to the ferrous iron atom in the direction opposite to this covalent link.
(e) **False** If the unfolding is effected with urea, then dialysis to remove the urea can reproduce the apoprotein. Addition of haem then restores myoglobin.

38 (a) **False** D-amino acids do not occur in haemoglobins.
(b) **True** Haemoglobin H, for example, has 4 β-chains.
(c) **True** An amino acid replacement often can alter the electric charge at that position.
(d) **True** Frequently this is the basis for the difference.
(e) **False** Hydroxylysine is an amino acid of the protein collagen.

39 The Michaelis constant (K_m) of an enzyme, catalysing a single substrate reaction, is:
(a) Approximately proportional to the velocity of the enzyme-catalysed reaction.
(b) Approximately proportional to the molar ratio of free to bound substrate under conditions when the proportion of enzyme sites occupied is small.
(c) Dependent on the enzyme concentration.
(d) Dependent on the substrate concentration.
(e) Dependent on the temperature.

40 The Michaelis constant (K_m) of an enzyme, catalysing a single substrate reaction, is:
(a) The equilibrium constant for the reaction between substrate and enzyme.
(b) An index of the catalytic power of the enzyme.
(c) A substrate concentration giving maximum reaction velocity.
(d) A substrate concentration giving half maximum reaction velocity.
(e) Only determinable if the pure enzyme is available.

41 The maximum velocity (V_{max}) of an enzyme-catalysed reaction (as usually defined and as used in the various forms of the Michaelis–Menten equation) is:
(a) A useful fundamental characteristic property of an enzyme.
(b) Only measurable provided the pure enzyme is available.
(c) Convertible into the turnover number of the enzyme if the enzyme molarity and the number of active catalytic sites per enzyme molecule are known.
(d) Convertible into the reaction velocity at a lower substrate concentration by multiplying V_{max} by the fraction of active sites filled at the lower substrate concentration.
(e) Often measured experimentally for the purpose of enzyme assay.

42 If reciprocals of reaction velocities (v^{-1}) are plotted against reciprocals of substrate concentrations $[S]^{-1}$ for an enzyme-catalysed reaction, a straight line is obtained that cuts the v^{-1} axis at 2.5×10^{-2} litre-minutes per micromole and the $[S]^{-1}$ axis at -2×10^{-2} litres per micromole. Which of the following statements are correct and which false?
(a) There can be no enzyme inhibitor present.
(b) The reaction is probably allosteric.
(c) The maximum velocity attained at high substrate concentrations (V_{max}) is 2.5×10^{-2} micromoles per litre per minute.
(d) Under the experimental conditions, K_m for the reaction is 50 micromoles per litre.
(e) A velocity of $V_{max}/4$ is given by a substrate concentration of approximately 16.7 micromoles per litre.

39 (a) **False** The velocity varies with both substrate and enzyme concentrations, whilst the constant K_m is independent of these.
(b) **True** $K_m = [E][S]/[ES]$ in the steady state. If relatively few sites are occupied $[E] \simeq [E_{TOTAL}]$, which is constant. Thus K_m is approximately proportional to the ratio $[S]/[ES]$ and therefore to the ratio of moles of free substrate to moles of bound substrate in the reaction medium.
(c) **False** Determinations of K_m with different enzyme concentrations should give the same value.
(d) **False**
(e) **True** $K_m = (k_2 + k_3)/k_1$ and each of these 3 velocity constants is a function of the absolute temperature.

40 (a) **False** $K_m = [E][S]/[ES]$ under *steady state*, not under *equilibrium*, conditions.
(b) **False** The K_m value is unrelated to the rate at which ES breaks down to give the reaction products.
(c) **False** Any sufficiently high substrate concentration will give a reaction velocity experimentally indistinguishable from V_{max}.
(d) **True** This is often given as the definition of K_m.
(e) **False** It is important to appreciate that enzyme-kinetics investigations require neither the pure enzyme nor information on its concentration in the catalytically active sample under investigation.

41 (a) **False** The magnitude of V_{max} depends on the enzyme concentration $[E_{TOTAL}]$.
(b) **False** An enzymically active sample is all that is required.
(c) **True** Turnover number = V_{max} divided by [(enzyme molarity) × (number of active sites per enzyme molecule)].
(d) **True**
(e) **True** Enzyme assays are determinations of enzyme catalytic activity, often measured at high substrate concentrations.

42 (a) **False** An inhibitor might or might not be present.
(b) **False**
(c) **False** $V_{max} = (2.5 \times 10^{-2})^{-1} = 40$ micromoles per litre per minute.
(d) **True** $K_m = -(-2 \times 10^{-2})^{-1} = 50$ micromoles per litre.
(e) **True** Establishing this requires use of the Michaelis–Menten equation:

$$V_{max}/v = 1 + K_m/[S]$$

Replacing v by $V_{max}/4$ gives $3 = K_m/[S]$ or $[S] = K_m/3 \simeq 16.7$ micromoles per litre.

43 The Lineweaver–Burk plot of v^{-1} *versus* $[S]^{-1}$ for an enzyme-catalysed reaction gives a straight line and in the presence of an inhibitor it gives a second straight line which intersects the first on the v^{-1} axis.
(a) The inhibition is competitive.
(b) The inhibition is irreversible.
(c) Substrate and inhibitor molecules both have affinity for the same enzyme sites.
(d) The Lineweaver–Burk line for the inhibited reaction lies below that of the uninhibited reaction for positive values of v^{-1} and $[S]^{-1}$.
(e) Lineweaver–Burk plots are important because they are the only way of obtaining linear (straight line) plots from reaction velocity and corresponding substrate concentration data for enzyme-catalysed reactions.

44 A non-competitive inhibitor differs from a competitive inhibitor in that it:
(a) Gives a straight line Lineweaver–Burk (v^{-1} against $[S]^{-1}$) plot, which intersects the corresponding line for the uninhibited reaction on $[S]^{-1}$ axis.
(b) Does not alter the measured K_m for the reaction.
(c) Is the same as an uncompetitive inhibitor.
(d) May attach itself to the enzyme at a site different from the active catalytic site.
(e) Reduces the affinity of substrate for enzyme.

45 The following reaction scheme represents the mechanism of the reaction that occurs when substrates S_1 and S_2 are allowed to react in the presence of enzyme (either E_1 or E_2) to give products P_1 and P_2.

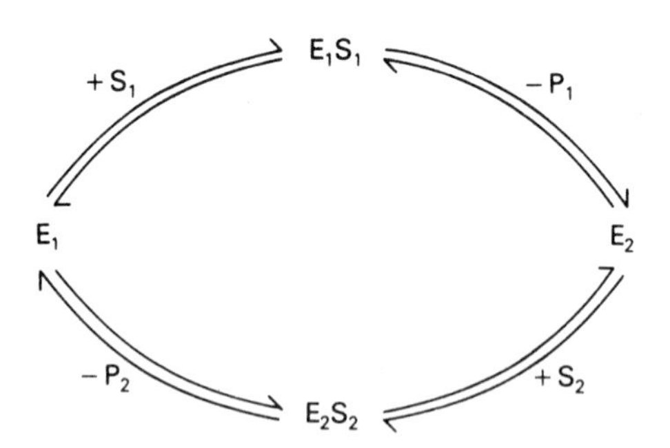

Which of the following are correct?
(a) E_1 and E_2 must have identical active sites.
(b) This mechanism is known as a ping-pong mechanism.
(c) Plotting v^{-1} against $[S_2]^{-1}$ will give a straight line at the steady state, if the velocities of the reactions of P_1 with E_2 and of P_2 with E_1 are both negligible.
(d) If P_1 is added to the initial reaction mixture it will act as a competitive inhibitor in relation to substrate S_1.
(e) If P_2 is added to the initial reaction mixture it will act as an uncompetitive inhibitor in relation to substrate S_2.

43 (a) **True** At sufficiently high substrate concentrations, both inhibited and uninhibited reactions have velocities that tend to the same V_{max} value.

(b) **False** Irreversible association with the inhibitor must reduce the V_{max} value.

(c) **True** Competition occurs for the available active sites.

(d) **False** Inhibition means that at any [S] value the value of v is reduced and thus v^{-1} is *raised.*

(e) **False** There are other ways of plotting to give straight lines. The Hofstee plot for example involves plotting v against v/[S].

44 (a) **True** The measured K_m is unaltered and consequently the intercepts of both lines on the $[S]^{-1}$ axis must be identical.

(b) **True**

(c) **False** An uncompetitive inhibitor does not attach itself to the same type of enzyme molecule that the substrate molecules interact with. Thus, in the case of a single substrate reaction, an uncompetitive inhibitor can only associate with the already formed enzyme–substrate complex.

(d) **True**

(e) **False** See note (a).

45 (a) **False** E_1 and E_2 will probably differ in at least one active site group (e.g. a group in the active site might be in an oxidized form in E_2 and in a reduced form in E_1).

(b) **True**

(c) **True** The steady-state kinetics will obey the equation

$$\frac{V_{max}}{v} = 1 + \frac{(K_m)_1}{[S_1]} + \frac{(K_m)_2}{[S_2]},$$

so that plotting v^{-1} against either $[S_1]^{-1}$ or against $[S_2]^{-1}$ will give straight lines.

(d) **False** E_1S_1 splits up to give E_2 and P_1, consequently in reverse, P_1 will attach itself to the active site of E_2. As S_1 only attaches itself to the active site of E_1, P_1 and S_1 do *not* compete for the *same* active site.

(e) **True** Substrate S_2 interacts with the active site of E_2, whilst P_2 attaches itself to the active site of E_1. Adding P_2 will inhibit the overall reaction (product inhibition), but as S_2 interacts with E_2 and P_2 does not, the inhibition is *uncompetitive.*

46 Citrate exerts effects on the activities of certain enzymes involved in glycolysis and in fatty acid synthesis.
(a) These effects are all examples of competitive inhibition.
(b) The effect on phosphofructokinase is an example of negative feedback control
(c) The effect on acetyl-CoA carboxylase is an example of negative feedback control.
(d) Effects on phosphofructokinase and on acetyl-CoA carboxylase are examples of allosteric effects.
(e) Both citrate and palmitoyl-CoA have similar effects on acetyl-CoA carboxylase.

47 The active site of chymotrypsin:
(a) Catalyses rapid hydrolysis of a peptide bond of type

$$—NH—CH_2—CO—NH—\underset{\underset{CH_2—C_6H_5}{|}}{CH}—CO—$$

(b) Contains a serine residue.
(c) Is uninfluenced in its activity by pH change.
(d) Is inactivated by di-isopropyl phosphofluoridate (DFP).
(e) Contains a histidine residue.

48 Human serum alkaline phosphatase levels can provide information relevant to:
(a) Obstruction of the bile duct.
(b) The progress of carcinoma of the prostate.
(c) Bone disease associated with osteoblastic activity.
(d) Osteolytic disease.
(e) Myocardial infarction.

49 Which of the following descriptions can be applied correctly to glucose 6-phosphate?
(a) It is a low energy phosphate ester.
(b) Its metabolic conversion never results in a net increase in fatty acid content.
(c) Metabolism can convert its carbon structure into the glycerol part of triglycerides.
(d) It is used to supply glucose units for glycogen synthesis.
(e) It reacts with phenols to form glucuronides.

46 (a) **False** The important effects are allosteric effects.
(b) **True** Phosphofructokinase is a key enzyme in glycolysis, allosterically influenced by various small molecules. In the case of citrate, we have a product of the pathway having an inhibiting effect on an enzyme early in the pathway.
(c) **False** The effect of citrate on acetyl-CoA carboxylase stimulates fatty acid synthesis. This is a case of a positive feedforward effect.
(d) **True**
(e) **False** Both are allosteric effects, but whilst the effect of citrate is activation, that of palmitoyl-CoA is inhibition.

47 (a) **False** Under appropriate conditions, chymotrypsin rapidly hydrolyses peptide bonds to give peptides in which the *new carboxyl* terminals are one or more of the amino acids: phenylalanine, tyrosine, and tryptophan.
(b) **True**
(c) **False** The optimum pH for activity is approximately 8·0.
(d) **True** DFP reacts with the active site serine residue as follows:

$$\begin{array}{l} | \\ \text{NH} \\ | \\ \text{CH—CH}_2\text{OH} \\ | \\ \text{CO} \\ | \end{array} + \begin{array}{c} \text{OCH(CH}_3)_2 \\ | \\ \text{F—P}^+\text{—O}^- \\ | \\ \text{OCH(CH}_3)_2 \end{array} \xrightarrow[-HF]{} \begin{array}{l} | \\ \text{NH} \\ | \\ \text{CH—CH}_2\text{O—} \\ | \\ \text{CO} \\ | \end{array}\begin{array}{c} \text{OCH(CH}_3)_2 \\ | \\ \text{P}^+\text{—O}^- \\ | \\ \text{OCH(CH}_3)_2 \end{array}$$

(e) **True**

48 (a) **True** Alkaline phosphatase is synthesized in the cells lining the bile canaliculi. Biliary obstruction causes regurgitation into the blood stream.
(b) **False** This information is provided by acid phosphatase determinations.
(c) **True** Examples are osteomalacia and rickets, and Paget's disease.
(d) **False**
(e) **False**

49 (a) **True** It is the phosphate ester of an alcohol with no structural features that would make it a high energy compound.
(b) **False** Fatty acids would be formed in the fed state.
(c) **True** The 6C unit of glucose is converted into two 3C compounds that can form the glycerol part of triglycerides.
(d) **True**
(e) **False** Glucuronides are formed from UDP-glucuronic acid.

50 Fluoroacetate, FCH_2COO^-, a rodenticide, inhibits the operation of the citric acid cycle because:
(a) It inactivates SH groups in enzymes.
(b) It prevents the formation of citrate by the condensing enzyme.
(c) Its structure is sufficiently like succinate to inhibit succinate dehydrogenase.
(d) Fluoride is a general enzyme poison.
(e) Fluorocitrate is a competitive inhibitor of aconitase.

51. The following statements relate to the citric acid cycle. Are they true or false?
(a) Acetyl-CoA and oxaloacetate react to form citrate.
(b) The cycle reforms oxaloacetate.
(c) CO_2 evolved in one turn of the cycle originates from the 2 C-atoms of the acetyl group in acetyl-CoA used to form citrate.
(d) It is involved in gluconeogenesis from glutamate.
(e) It is the main source of NADPH for fatty acid synthesis.

52 Which of the following statements apply to β-oxidation of fatty acids?
(a) Free acids must be esterified with coenzyme A.
(b) The process involves the reduction of $NADP^+$.
(c) β-oxidation take place in the mitochondria.
(d) The carbon atoms removed during oxidation are available for further metabolism.
(e) Continued β-oxidation is dependent on a supply of ADP.

53 Which of the following correctly apply to the metabolic 'fed state' when glucose is being stored as glycogen and triglyceride?
(a) Glycogen is formed through α-1,4 links between glucose molecules extending the polysaccharide chain at its reducing end.
(b) Each 2 carbon step in chain elongation of fatty acids between C_2 and C_{16} acids involves malonyl-CoA.
(c) Fatty acid synthesis obtains necessary reducing power from the pentose phosphate pathway.
(d) UDP-glucose is the glucose-transferring substrate in glycogen synthesis.
(e) The essential fatty acids, linoleic and arachidonic acids are synthesized from glucose.

54 Which of the following are characteristic of the fasting metabolic state?
(a) Enhanced lipolysis occurs in adipose tissue.
(b) Ketogenesis is enhanced in the liver.
(c) Gluconeogenesis occurs in the liver.
(d) Insulin output increases.
(e) Glucose is formed from muscle glycogen *via* glucose 6-phosphate.

55 Which of the following has one or more of the B-group vitamins built into its structure?
(a) Thiamin pyrophosphate.
(b) UDP-glucuronic acid.
(c) NAD^+.
(d) The flavoprotein, succinate dehydrogenase.
(e) Glucose 6-phosphate.

50 (a) **False** This applies to iodoacetate.
(b) **False** Fluoroacetyl-CoA does not prevent the formation of citrate.
(c) **False** It is not very similar structurally to succinate.
(d) **False**
(e) **True** Fluoracetate is metabolized to fluorocitrate, which then blocks the formation of *cis*-aconitate by aconitase.

51 (a) **True**
(b) **True**
(c) **False** It comes from 2 C-atoms in the oxaloacetate.
(d) **True** Transamination converts glutamate into α-oxoglutarate, which the cycle converts into oxaloacetate.
(e) **False** The source of NADPH for fatty acid synthesis is the pentose phosphate pathway.

52 (a) **True** The formation of the long-chain acyl-CoA derivatives is the essential preliminary for fatty acid oxidation.
(b) **False** It involves the reduction of NAD^+.
(c) **True**
(d) **True** They form the acetyl part of acetyl-CoA.
(e) **False** It is dependent on ATP, NAD^+ and HS(CoA).

53 (a) **False** It takes place at the non-reducing end of the existing unit.
(b) **True**
(c) **True** NADPH formed in the pentose phosphate pathway provides this reducing power.
(d) **True**
(e) **False** They must be supplied in the diet.

54 (a) **True** Mediated by the fasting state hormones.
(b) **True** Ketones can be used as an energy source.
(c) **True** Glucose is synthesized from amino acids.
(d) **False**
(e) **False** Muscle lacks glucose 6-phosphatase.

55 (a) **True** Thiamin is vitamin B_1.
(b) **False**
(c) **True** Nicotinamide is part of the NAD^+ structure. Niacin is nicotinic acid.
(d) **True** Riboflavin is part of the flavoprotein structure.
(e) **False**

56 Many metabolic transformations involve the oxidation or reduction of nicotinamide adenine dinucleotides. Which of the following statements are true and which are false?

(a) Conversion of dihydroxyacetone phosphate to α-glycerophosphate involves the oxidation of NADPH.
(b) Conversion of glucose 6-phosphate to 6-phosphogluconate involves the reduction of $NADP^+$.
(c) Synthesis of long-chain fatty acids from acetyl-CoA involves the oxidation of NADPH.
(d) Formation of fumarate from succinate involves the oxidation of NADH.
(e) Conversion of pyruvate to acetyl groups in acetyl-CoA involves the reduction of NAD^+.

57 Consider the processes involved in the oxidation of H atoms (combined in metabolites) to give water. Which of the following statements are true and which are false?

(a) Proteins with conjugated haem groups, called cytochromes, are involved.
(b) The rate of electron transport is increased when the ATP/ADP ratio is high.
(c) Oxidation of the 2 H atoms from the conversion of succinate to fumarate can produce, as a consequence, 2 ATP molecules.
(d) If phosphorylation is uncoupled, electron transport can never occur.
(e) Oxidative phosphorylation occurs in the cytosol.

58 Which of the following might apply to a patient with an inadequately controlled insulin-dependent diabetes?

(a) An abnormally slow drop in blood glucose concentration following a meal.
(b) An abnormally low concentration of fatty acids in the blood.
(c) A high concentration of ketones in the blood.
(d) Glycosuria.
(e) An abnormally high rate of glycogen synthesis.

59 Which of the following might apply to an infant with glycogen storage disease, type I (von Gierke's disease)?

(a) Blood glucose is abnormally low after a 6-hour fast.
(b) Extensive deposition of glycogen occurs in the liver.
(c) Liver glucose 6-phosphatase activity is abnormally high.
(d) Blood lactic acid levels are high.
(e) Hypolipaemia occurs.

56 (a) **False** It involves the oxidation of NADH.
(b) **True**
(c) **True** Reducing power is provided by NADPH.
(d) **False** it involves reduction of FAD.
(e) **True**

57 (a) **True**
(b) **False** The opposite is true in a coupled system.
(c) **True** This is in contrast to the 3 ATP molecules produced by the oxidation of the 2 H atoms resulting from the conversion:

$$NADH + H^+ \longrightarrow NAD^+ + 2H$$

(d) **False** Electron transport can occur irrespective of the accompanying conversion of ADP to ATP.
(e) **False** These are mitochondrial processes.

58 (a) **True** This is a consequence of low insulin levels.
(b) **False** Increased liberation of fatty acids into the blood stream occurs and this leads to a hyperlipaemia.
(c) **True** This is due to increased fat metabolism.
(d) **True** Increased concentration of glucose in blood *may* give rise to glycosuria.
(e) **False**

59 (a) **True** This condition is due to an inborn deficiency of glucose 6-phosphatase. Thus, glucose 6-phosphate is not hydrolysed effectively to give free glucose.
(b) **True** Even in the fasting state, very active glycogenesis occurs.
(c) **False** See (a) above.
(d) **True** The glucose 6-phosphatase deficiency results in the build up of intermediates in the glycolytic and gluconeogenic pathways and in particular, as a result, blood lactic acid concentration is raised.
(e) **False** Prolonged hypoglycaemia causes secondary disturbances of lipid metabolism with excessive mobilization of fat from adipose tissue. A net consequence is the development of hyperlipaemia.

60 In the urea cycle, the enzyme ornithine transcarbamylase is concerned with:
(a) The formation of citrulline from ornithine.
(b) The formation of ornithine from citrulline.
(c) The formation of urea from arginine.
(d) The transamination of ornithine.
(e) The hydrolysis of ornithine.

61 The following statements apply to the biochemical disorder associated with the syndrome called argininosuccinic aciduria.
(a) There is a lack of the enzyme carbamyl phosphate synthetase.
(b) Argininosuccinate accumulates in blood and urine.
(c) There is a lack of argininosuccinase activity.
(d) A normal amount of urea is produced.
(e) There is a lack of the enzyme arginase.

62 If 5 g of ammonium chloride were administered to a subject, which of the following are likely consequences?
(a) Urinary pH rises sharply.
(b) Some increase in output of urine occurs.
(c) Glutaminase activity increases in kidney tubule cells.
(d) Glutamate decarboxylase activity increases.
(e) Reducing substances appear in the urine.

63 Which of the following statements are true and which false for the compound creatinine?
(a) The daily urinary output of creatinine for any particular normal adult is approximately constant.
(b) The weight excreted per day depends mainly on the body's muscle mass.
(c) It is the metabolic precursor of muscle creatine.
(d) The molecule acquires its methyl group from the amino acid methionine.
(e) It is formed by the decarboxylation of histidine.

60 (a) **True** Citrulline is formed from ornithine and carbamyl phosphate.

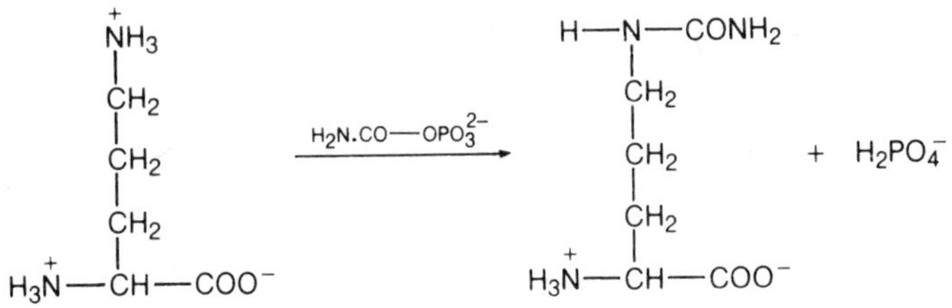

$$\begin{array}{c}\overset{+}{N}H_3\\ |\\ CH_2\\ |\\ CH_2\\ |\\ CH_2\\ |\\ H_3\overset{+}{N}-CH-COO^-\end{array} \xrightarrow{H_2N.CO-OPO_3^{2-}} \begin{array}{c}H-N-CONH_2\\ |\\ CH_2\\ |\\ CH_2\\ |\\ CH_2\\ |\\ H_3\overset{+}{N}-CH-COO^-\end{array} + H_2PO_4^-$$

(b) **False** Citrulline is converted into ornithine in stages, *via* arginosuccinate and arginine.
(c) **False** Arginase converts arginine into urea.
(d) **False**
(e) **False**

61 (a) **False** This would apply to the syndrome ammonaemia.
(b) **True**
(c) **True** This is the reason for the accumulation of arginosuccinate in (b).
(d) **True** When the concentration of substrate arginosuccinate is high enough, then the formation of arginine and consequently of urea can proceed at a normal rate.
(e) **False**

62 (a) **False** The urinary pH will fall.
(b) **True** Chloride output will increase, together with compensating quantities of anions including Na^+. As a consequence some measure of diuresis occurs.
(c) **True** Ingested NH_4^+ is largely converted into urea in the liver and the H^+ ions released from the NH_4^+ are sequestered by the blood buffer system and this 'acidity' is then removed by the kidneys. In renal tubule cells, glutaminase activity increases, glutamine is hydrolysed to give NH_3, and this combines with H^+ ions to give NH_4^+ ions, which are excreted.
(d) **False** Glutamate decarboxylase forms γ-aminobutyric acid from glutamate.
(e) **False**

63 (a) **True** As a consequence, the determination of urinary creatinine may be used as a check on the accuracy of collection of 24-hour urine specimens.
(b) **True**
(c) **False** It is the product of creatine metabolism.
(d) **True** Guanidylacetate is methylated by S-adenosyl-methionine (+ ATP) to give creatine phosphate, from which creatinine is produced.
(e) **False** Although the 5-membered ring of creatinine is similar to that of histidine, the creatinine ring structure originates from the amino acids arginine and glycine and not from modification of the 5-membered ring of histidine.

64 Mammalian biosynthesis of the following require one or more methyl groups to be provided by transfer from methionine through the intermediate S-adenosyl-methionine.
(a) Valine.
(b) Uridine monophosphate.
(c) Choline.
(d) Adrenaline.
(e) Ribosomal RNA.

65 Pyridoxal phosphate is required as a cofactor in the following reactions of particular amino acids:
(a) Deamination.
(b) Decarboxylation.
(c) Transamination.
(d) Desulphuration.
(e) Acetylation.

64 (a) **False**
(b) **False**
(c) **True** Serine is decarboxylated to give ethanolamine, which is then progressively methylated at the amino group to give choline:

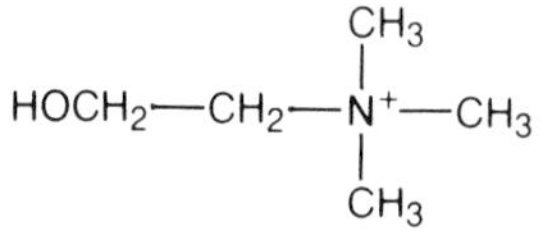

$$HOCH_2—CH_2—\overset{\displaystyle CH_3}{\underset{\displaystyle CH_3}{N^+}}—CH_3$$

(d) **True** Noradrenaline is methylated to give adrenaline.

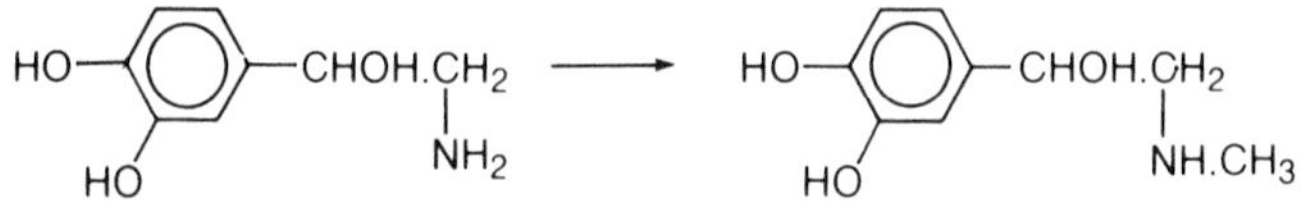

$$\text{HO(HO)}C_6H_3—CHOH.CH_2NH_2 \longrightarrow \text{HO(HO)}C_6H_3—CHOH.CH_2NH.CH_3$$

(e) **True** In the formation of rRNA in mammalian cells, transcription gives a precursor RNA, which is methylated by S-adenosyl-methionine and then cleaved in several steps to give the 28S and 18S RNA structures, which are ribosomal components. In the methylation process, about 1% of the ribose rings in the 45S precursor are methylated, the methylation occurring at the 2′-position.

65 (a) **True** It is a co-substrate in the non-oxidative deamination of serine and of threonine, e.g.

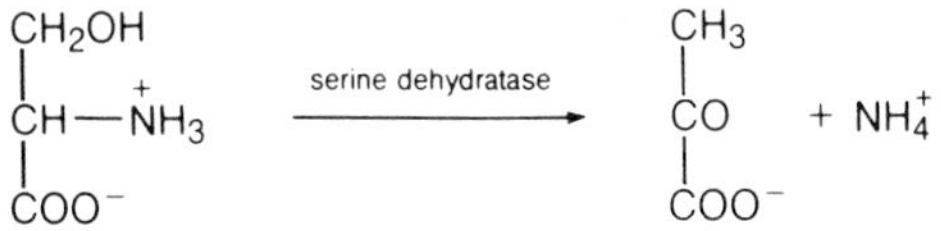

$$\begin{matrix} CH_2OH \\ | \\ CH—\overset{+}{N}H_3 \\ | \\ COO^- \end{matrix} \xrightarrow{\text{serine dehydratase}} \begin{matrix} CH_3 \\ | \\ CO \\ | \\ COO^- \end{matrix} + NH_4^+$$

(b) **True** It acts as a co-decarboxylase in the decarboxylation of tyrosine, arginine and glutamic acid.
(c) **True** It is required for transamination reactions between amino acids and oxo acids.
(d) **True** Pyridoxal phosphate is a co-substrate in desulphuration reactions of cysteine and of homocysteine, e.g.

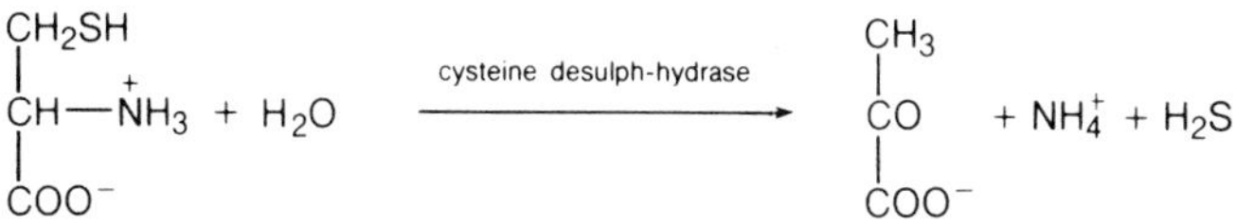

$$\begin{matrix} CH_2SH \\ | \\ CH—\overset{+}{N}H_3 \\ | \\ COO^- \end{matrix} + H_2O \xrightarrow{\text{cysteine desulph-hydrase}} \begin{matrix} CH_3 \\ | \\ CO \\ | \\ COO^- \end{matrix} + NH_4^+ + H_2S$$

(e) **False**

66 The following reactions will be catalysed by a suitable liver transaminase:

(a) Oxidative deamination of glutamate producing oxoglutarate.
(b) Transfer of an amino group from glycine to arginine.
(c) Transfer of an amino group from glutamate to pyridoxal phosphate bound to the enzyme.
(d) Transfer of an amino group from (enzyme-bound) pyridoxamine phosphate to pyruvate.
(e) Transfer of a phosphate group from pyridoxal phosphate to pyridoxamine.

66 (a) **False**
(b) **False**
(c) **True** The reaction occurring at the active site of the enzyme is of the following type:

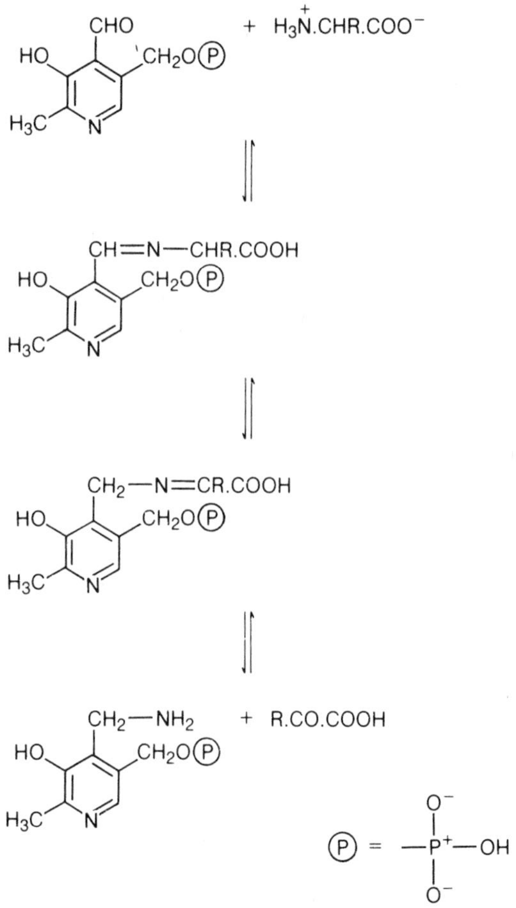

CHO, HO, CH_2OⓅ, H_3C, N + $H_3\overset{+}{N}.CHR.COO^-$

⇌

CH=N—CHR.COOH, HO, CH_2OⓅ, H_3C, N

⇌

CH_2—N=CR.COOH, HO, CH_2OⓅ, H_3C, N

⇌

CH_2—NH_2, HO, CH_2OⓅ, H_3C, N + R.CO.COOH

Ⓟ = $-\overset{O^-}{\underset{O^-}{P^+}}-OH$

In the full transamination process, whilst one amino acid is being converted into the oxo acid, another oxo acid is being simultaneously converted into its corresponding amino acid.
(d) **True** Refer to (c) above.
(e) **False**

67 Which of the following statements are true, and which are false for the condition known as phenylketonuria?
(a) It can only be diagnosed after the first year of life.
(b) The enzyme phenylalanine hydroxylase is defective.
(c) Tyrosine builds up in the blood.
(d) High levels of phenylpyruvic acid are excreted in the urine.
(e) Blood from phenylketonurics contains high levels of phenylalanine.

68 In animals the following compounds are mainly formed by biosynthesis from tyrosine:
(a) Serotonin.
(b) Thyroxine.
(c) Phenylalanine.
(d) Adrenaline.
(e) Homogentisic acid.

69 Consider the substance L-3,4-dihydroxyphenylalanine (dopa):

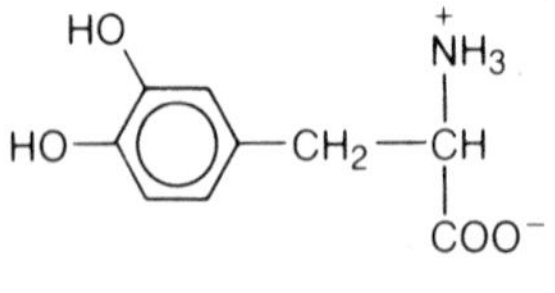

HO
NH_3^+
HO— (benzene ring) —CH_2—CH
COO^-

(a) It can be formed from L-tyrosine.
(b) It is a major precursor of tyrosine in the body.
(c) L-3,4-dihydroxyphenylethylamine (dopamine) is formed by decarboxylation of dopa.
(d) Dopa is an important neurotransmitter.
(e) Dopa is an intermediate compound in the biosynthesis of adrenaline.

67 (a) **False** It is diagnosed soon after birth.
(b) **True** Phenylalanine hydroxylase normally converts phenylalanine to tyrosine.
(c) **False** The synthesis of tyrosine from phenylalanine is defective (see (b)).
(d) **True** Phenylpyruvic acid is formed from phenylalanine by transamination.
(e) **True** Phenylalanine concentration is markedly increased in all body fluids. Metabolites: phenylpyruvic acid, phenyl-lactic acid, phenylacetic acid and phenylacetylglutamine also appear in the urine.

68 (a) **False** Serotonin is 5-hydroxytryptamine.
(b) **True** Tyrosine is a precursor of the hormone thyroxine

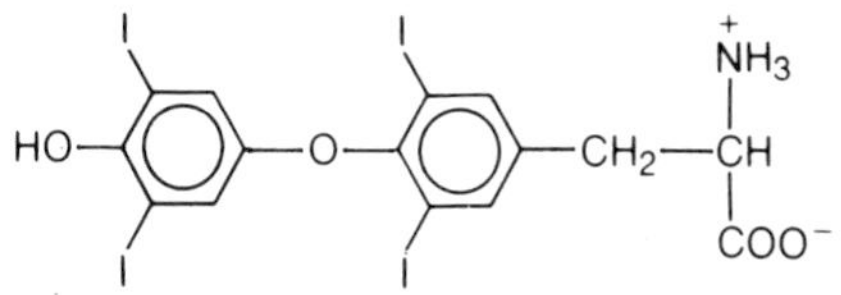

(c) **False** The hydroxylation of phenylalanine to form tyrosine is effected by a mixed function oxidase. Phenylalanine is oxidized to tyrosine and simultaneously tetrahydrobiopterin is oxidized to the dihydro-derivative, the oxidant being molecular oxygen. The dihydro-derivative is reduced by NADPH. The overall process is effectively irreversible.
(d) **True** Tyrosine is further hydroxylated to give dopa, which is decarboxylated and the dopamine formed is hydroxylated at the α-carbon atom. The noradrenaline produced forms adrenaline on methylation.
(e) **True** Homogentisic acid is a component of the pathway of tyrosine catabolism. Transamination of tyrosine is followed by an oxidative decarboxylation to yield homogentisic acid. Further oxidation opens the benzene ring.

69 (a) **True** Hydroxylation of tyrosine by tyrosine hydroxylase gives dopa.
(b) **False**
(c) **True** The enzyme dopa decarboxylase converts dopa into dopamine.
(d) **False** It is an important precursor, but is not considered a transmitter in itself.
(e) **True**

70 Which of the following statements are true and which false for the condition known as alcaptonuria?
(a) The patient's urine darkens on exposure to air.
(b) There is a lack of the enzyme phenylalanine hydroxylase.
(c) There is a lack of homogentisic acid oxidase.
(d) There is no tyrosinase activity.
(e) A generalized amino aciduria exists.

71 Which of the following statements are correct and which are false for the amino acid γ-aminobutyric acid?
(a) It inhibits anticholinesterase activity.
(b) It is an inhibitory neurotransmitter substance.
(c) It is synthesized in the brain from glutamate.
(d) The enzyme glutamate dehydrogenase is directly responsible for its synthesis.
(e) It is an essential amino acid.

72 Which of the following correctly apply to the *dietary essential amino acids* of the normal human adult and which do not?
(a) They are only used for the synthesis of important body proteins.
(b) If any one is consumed in great quantity with an otherwise balanced diet, a positive nitrogen balance will develop.
(c) Their carbon skeletons and groups cannot be synthesized by humans in the required quantities.
(d) Consumption of a mixture of all the essential amino acids in sufficient quantities could allow the body to make all the proteins it needs (assuming that other minor factors are available.)
(e) None is a substrate for a transaminase enzyme.

70 (a) **True** Exposed to oxygen, the pigment alcapton forms in the urine.
(b) **False** Phenylalanine hydroxylase is defective in phenylketonuria.
(c) **True**
(d) **False** Tyrosinase activity is absent in the condition known as albinism.
(e) **False**

71 (a) **False**
(b) **True** Interference with its synthesis is believed to give rise to convulsions.
(c) **True** It is formed by decarboxylation of glutamate.
(d) **False** It is formed from glutamate by the enzyme glutamate decarboxylase.
(e) **False**

72 (a) **False** They have important special functions in addition to their role in protein structure.
(b) **False**
(c) **True** This is characteristic of essential amino acids.
(d) **True**
(e) **False**

73 The graph shows the daily nitrogen balance of a human adult fed on an adequate synthetic diet in which a mixture of pure amino acids, together with the minimum requirement of vitamins, provided the sole sources of nitrogen.

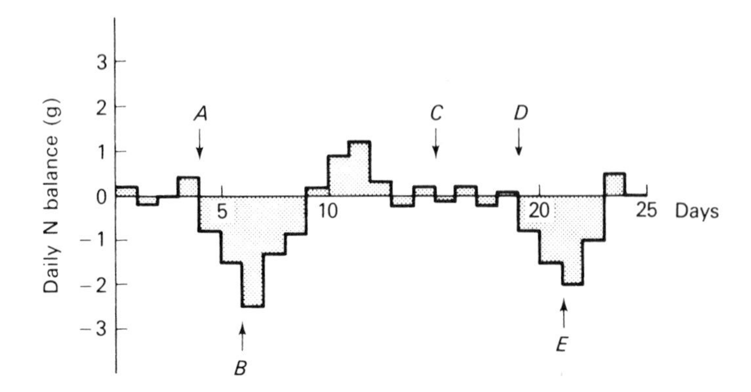

Decide which of the following are consistent, and which are not, with the results portrayed:

(a) Alanine was omitted from the diet at *A* and restored at *B*.
(b) Glycine was omitted from the diet at *C*.
(c) Tryptophan intake was reduced at *D*.
(d) Tryptophan intake was reduced at *D* and phenylalanine added at *E*.
(e) Leucine intake was reduced at *D* and replaced by isoleucine at *E*.

74 Which of the following statements concerning nucleosides and nucleotides are true and which false?

(a) Nucleosides and deoxynucleosides are similar in that they both possess two hydroxyl groups attached to the oxygen-containing ring.
(b) A nucleotide, when chemically hydrolysed under mild alkaline conditions gives a nucleoside and phosphate, but when chemically hydrolysed under mild acidic conditions gives a sugar phosphate and a base.
(c) Adenine nucleotides are present as components of the molecules of coenzyme A, NAD^+ and FAD.
(d) Polynucleotides can be formed from nucleoside triphosphates.
(e) A deoxynucleoside molecule must contain one of the four bases that are present in DNA.

75 In relation to nucleic acids:

(a) Base pairing occurs between a pyrimidine base and a purine base.
(b) The bonding between a guanine base and a cytosine base occurs with the formation of two hydrogen bonds.
(c) The two polynucleotide chains in DNA have opposite polarity.
(d) A hydrogen-bonded base pair in a DNA double helix forms an approximately planar structure.
(e) Base pairing is markedly less effective when thymine is the base rather than uracil because of interference to hydrogen bonding produced by the 5-methyl group.

73 (a) **False** Alanine is not an essential amino acid.
(b) **True** Glycine is not an essential amino acid and thus its removal at point *C* will make no permanent effect on the N-balance.
(c) **True** Tryptophan is an essential amino acid and its removal at point *D* is consistent with a negative N-balance.
(d) **False** Tryptophan intake was reduced at *D* giving a negative N-balance but this could not be restored at *E* by phenylalanine.
(e) **False** Leucine deficiency cannot be replaced by isoleucine.

74 (a) **False** Whereas nucleosides possess hydroxy groups at the 2′ and 3′-positions, the biologically important deoxynucleosides are 2′-deoxy compounds and lack the 2′-hydroxy group.
(b) **True**
(c) **True**
(d) **True**
(e) **False** Other bases might be present, e.g. 2′-deoxyuridine contains uracil.

75 (a) **True** Apart from other considerations, the invariable pairing of a pyrimidine with a purine preserves the correct geometry of the double helix.
(b) **False** G≡≡≡C bonding involves three hydrogen bonds per base pair.
(c) **True** This term refers particularly to polarity of direction. Thus the two H-bonded chains in DNA are of type:
5′———— TATG ————3′
3′———— ATAC ————5′
(d) **True** This enables base pairs to be 'stacked' on top of each other in the axial region of the DNA double helix.
(e) **False** The 5-methyl group of thymine is sufficiently far away not to interfere with the hydrogen bonding.

76 In relation to particular DNAs:
- (a) *E. coli* contains double-stranded circular DNA.
- (b) Smallpox and herpes simplex viruses are examples of DNA viruses.
- (c) Influenza A virus and polioviruses are examples of DNA viruses.
- (d) An appropriate DNA polymerase and DNA ligase are required for the *in vitro* synthesis of some bacteriophages.
- (e) All oncogenic viruses are DNA viruses.

77 Consider the synthetic polynucleotide known as poly-U (i.e., polyuridylic acid).
- (a) Complete hydrolysis gives uracil, ribose, and phosphate.
- (b) It migrates to the positive electrode on electrophoresis at pH 8.
- (c) When used as a synthetic messenger for polypeptide synthesis in a ribosomal preparation, only one type of amino acid is incorporated into the peptide.
- (d) It forms a double helix by base pairing with itself.
- (e) It acts as a transfer-RNA for phenylalanine.

78 Which of the following are true and which are false, when applied to protein biosynthesis on ribosomes:
- (a) Each transfer-RNA 'recognizes' a triplet codon on messenger-RNA in order to position correctly the amino acid that it carries.
- (b) Each amino acid is directly attached to three nucleotides specific for that amino acid.
- (c) Each transfer-RNA (usually loaded with its appropriate amino acid) has a sequence of three nucleotides (in a non-hydrogen-bonded loop) which base pairs only with an appropriate codon.
- (d) Three adjacent bases must be changed on messenger-RNA before the synthesized protein will have one of its amino acids replaced by another.
- (e) Messenger-RNA is read from the 5′ end toward the 3′ end, corresponding to the synthesis of protein from the C-terminus toward the N-terminus.

79 Considering the processes involved in protein synthesis, which of the following statements are true and which are false?
- (a) An amino acid molecule is linked to a transfer-RNA molecule by an ester linkage at the 2′-hydroxy group of the ribose part of the 3′-terminal nucleotide.
- (b) In the synthesis of many protein molecules, a methionine residue is removed at the N-terminal end before synthesis of the polypeptide chain is complete.
- (c) Messenger-RNA is translated from the 3′ end towards the 5′ end.
- (d) A nascent polypeptide chain is associated with the larger of the two ribosomal sub-units.
- (e) The messenger-RNA for secretory proteins usually has at the 5′ end, a base sequence coding for a precursor (signal) peptide.

76 (a) **True** The term 'circular' means non-terminating chains. Both genetic studies and electron microscopy have indicated that *E. coli* has a circular chromosome.
(b) **True**
(c) **False** Influenza and polioviruses are RNA viruses.
(d) **True** DNA polymerases catalyse extension of DNA chains in the 5′ → 3′ direction. DNA ligases catalyse the linking of two DNA chains, a phosphodiester bond being formed.
(e) **False** Many oncornaviruses are known, their replication involves the action of a reverse transcriptase enzyme.

77 (a) **True**
(b) **True** Ionized phosphate groups will produce migration to a positive electrode.
(c) **True** Polyphenylalanine will be formed.
(d) **False** U≡≡≡U pairing has extremely low probability.
(e) **False** It acts as a messenger-RNA for phenylalanine.

78 (a) **True** A specific amino acid has to be located at a particular locus and because of this, each amino acid–transfer-RNA complex recognizes a triplet codon on messenger-RNA.
(b) **False** Each amino acid is attached to the adenylic acid of a terminating sequence ———CCA, which is common to different transfer-RNA molecules.
(c) **True**
(d) **False** A single point mutation in the DNA, resulting in only one base being changed in the messenger RNA, may be sufficient to replace one amino acid by another. This is the case, for example, with HbA and HbS.
(e) **False** The protein is synthesized from the N-terminus to the C-terminus.

79 (a) **False** The linkage is at the 3′-hydroxy group of the terminal adenylic acid.
(b) **True** Thus in eukaryotes, AUG codes for methionine as an initiator of peptide chain synthesis, but the methionine is removed before synthesis of the chain is complete, unless it is the required N-terminal amino acid for the polypeptide.
(c) **False** Translation occurs in the 5′———→ 3′ direction.
(d) **True**
(e) **True** Following the AUG initiator codon is a group of signal codons, coding for a signal peptide.

80 During the translation of messenger-RNA, certain L-amino acids can be directly incorporated into polypeptide chains. These amino acids include the L-isomers of
(a) Asparagine.
(b) 4-Hydroxyproline.
(c) *p*-Hydroxyphenylalanine.
(d) Homoserine.
(e) Phosphoserine.

81 The following are codons for the translation of messenger-RNA into an amino acid sequence in a preparation of ribosomes, containing a transfer-RNA fraction, amino acids, Mg^{2+} ions and the minimum of other necessary components:

AAA = Lys, AAU = Asn, AUA = Ile,
UUA = Leu, UAU = Tyr, UUU = Phe.

An artificial polyribonucleotide was found to act as a template for the synthesis of a *single* polypeptide with a repeating structure:

——Ile–Tyr–Ile–Tyr——

This artificial polyribonucleotide could have been
(a) Poly-(AUUA).
(b) Poly-(AUAU).
(c) Poly-(UAU).
(d) Poly-(AUA).
(e) Poly-(UA).

82 Which statements are true and which false concerning the structure of lipid molecules:
(a) Lecithins are diglycerides with the remaining $-CH_2OH$ group of the glycerol esterified with choline phosphate.
(b) Phosphatidyl ethanolamines and phosphatidyl serines differ from corresponding lecithins in that the choline is replaced in the structure by ethanolamine and by serine respectively.
(c) Sphingomyelins and cerebrosides are other esters of glycerol.
(d) Cholesterol is partly esterified and is present in plasma β-lipoprotein as both cholesterol and as cholesteryl ester.
(e) Prostaglandins are steroid structures.

80 (a) **True** Asparagine (Asn) is directly incorporated.
(b) **False** Although 4-hydroxyproline residues are present in collagen, they are formed from proline residues already incorporated into polypeptide chains.
(c) **True** *p*-Hydroxyphenylalanine is the amino acid Tyrosine (Tyr).
(d) **False** Homoserine is involved in threonine, aspartate and methionine metabolism, but it is not a component of protein chains.
(e) **False** Phosphoserine residues are formed from serine residues.

81 (a) **False**
(b) **True** Note that if the Mg^{2+} concentration is sufficiently high in artificial (*in vitro*) systems, then a special initiator codon may not be required for the translation to occur.
(c) **False**
(d) **False**
(e) **True**

82 (a) **True** The general lecithin structure is

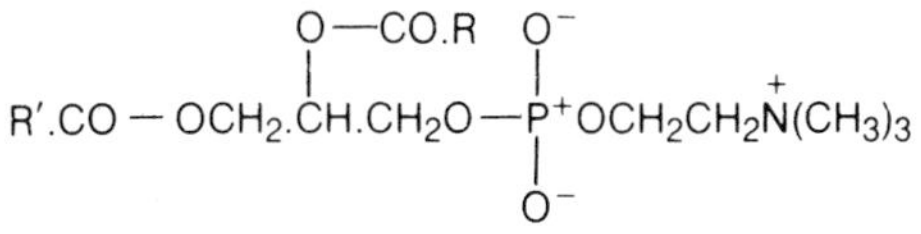

$$\begin{array}{c} \text{O—CO.R} \quad \text{O}^- \\ | \qquad\qquad | \\ \text{R}'\text{.CO} - \text{OCH}_2\text{.CH.CH}_2\text{O} - \text{P}^+\text{OCH}_2\text{CH}_2\overset{+}{\text{N}}(\text{CH}_3)_3 \\ | \\ \text{O}^- \end{array}$$

R and R′ being hydrocarbon chains.
(b) **True**
(c) **False** Sphyngomyelins and cerebrosides are derivatives, not of glycerol, but of the base sphingosine:

$$CH_3.(CH_2)_{12}.CH{=}CH.CHOH.CH(NH_2).CH_2OH$$

(d) **True**
(e) **False** Prostaglandins are related to unsaturated fatty acids, being specific oxidized derivatives of these acids.

83 The following statements concern lipids and lipoproteins.
(a) Niemann–Pick's, Gaucher's, and Tay–Sachs' diseases are examples of lipidoses arising from congenital deficiency in specific lysosomal enzymes concerned with the hydrolysis of particular complex lipids.
(b) Lipoprotein lipase differs from pancreatic lipase in acting on lipoprotein-bound triglyceride rather than on free triglyceride.
(c) Lipoprotein lipase levels in plasma drop dramatically after intravenous administration of heparin.
(d) Hyperlipidaemia type 1 is due to deficient tissue lipoprotein lipase activity giving high levels of chylomicrons in plasma.
(e) The α-lipoprotein electrophoretic fraction is rich in protein, whereas, by contrast, the β-lipoprotein fraction is rich in cholesterol.

84 Which of the following statements can be correctly applied to cholesterol and which cannot?
(a) It is a major constituent of plants, fungi, yeasts and bacteria.
(b) It is synthesized in mammals from acetyl-CoA.
(c) It is a component of many mucopolysaccharides.
(d) It is a constituent of many lipoproteins.
(e) Bile acids and sex hormones are synthesized from cholesterol in the body.

85 Aldosterone
(a) Is another name for the antidiuretic hormone.
(b) Increases the retention of sodium ions.
(c) Increases the retention of potassium ions.
(d) Is a glycoprotein.
(e) Is produced in the adrenal cortex.

86 A 24-hour urine collection from an adult man had a volume of 2 litres. A sample was diluted 100-fold. 2 ml of the diluted sample was mixed with 8 ml of picric acid reagent and 10 ml of M NaOH. The orange-red colour produced (specific for creatinine) had exactly the same absorbance at 520 nm as that of a solution prepared from 1 ml of a standard creatinine solution (10 mg/l) mixed with 4 ml of picric acid reagent and 5 ml of M NaOH.

What was the total creatinine content of the 24-hour urine sample?
(a) 20 mg.
(b) 500 mg.
(c) 1 g.
(d) 2 g.
(e) None of the above.

83 (a) **True** For example, sphingomyelinase is defective in Niemann–Pick disease. This enzyme removes, by hydrolysis, the choline phosphate group in sphingomyelin.
(b) **True**
(c) **False** Lipoprotein lipase is an enzyme found attached to the surface membranes of many tissues, but is not found in the blood of fasting normal subjects. Heparin frees the tissue-bound enzyme and enables it to pass into the blood.
(d) **True** Other features of this disease are enlarged liver and spleen, yellow areas (xanthomas) on the skin, and abdominal pain.
(e) **True** α-Lipoprotein contains about 50% protein, whilst β-lipoprotein contains about 45% total cholesterol. Other components have smaller proportions.

84 (a) **False** It is probably absent from most of these.
(b) **True** The synthesis proceeds *via* β-hydroxy-β-methyl-glutaryl-CoA, mevalonic acid, isopentenyl pyrophosphate, farnesyl pyrophosphate, and squalene.
(c) **False**
(d) **True** Both free cholesterol and its esters with fatty acids are present in lipoprotein fractions.
(e) **True** Bile acids are synthesized in the liver, androgens and oestrogens in the adrenal cortex. Steroid hormones are also produced in the gonads.

85 (a) **False** Aldosterone is a steroid that stimulates the retention of sodium ions and the excretion of potassium ions by the kidney. Antidiuretic hormone is a peptide (vasopressin) that effects renal reabsorption of water (facultative reabsorption).
(b) **True**
(c) **False**
(d) **False** It is a steroid.
(e) **True** It is synthesized from cholesterol.

86 (a) **False**
(b) **False**
(c) **False**
(d) **True** Standard solution 10 mg/l $\equiv$ 10 μg/ml. Hence test sample must contain 10 μg/ml creatinine. But 2 ml were used, thus total amount = 20 μg. Then amount before dilution = 2000 μg = 2 mg and amount in 2 litres = 2 g.
(e) **False**

87 Consider the Ames Clinistix glucose oxidase test for urinary glucose.

Which of the following solutions would give a positive reaction and which would not?

(a) 1% glucose in M hydrochloric acid.
(b) 1% lactose at pH 7.
(c) The mixed products of the enzymic hydrolysis of sucrose.
(d) Very dilute hydrogen peroxide.
(e) Acetoacetate.

88 A positive reaction was found when a patient's urine was tested with the Ames Phenistix.

Which of the following statements might explain the result and which would not?

(a) The patient is suffering from phenylketonuria.
(b) The patient has kidney damage.
(c) The patient took aspirin about 2 hours before collecting the urine sample.
(d) Reducing substances are present in the urine.
(e) The patient has been taking a cough mixture containing phenolic substances.

89 Which of the following statements apply to the Ames Albustix for urine testing and which do not?

(a) If protein is present, the bromophenol blue indicator changes colour due to a change in pH.
(b) When protein is present, it changes the ionization properties of the bromophenol blue indicator.
(c) Albustix are buffered and no significant pH change occurs.
(d) If protein is present, it acts as a buffer to the bromophenol blue indicator.
(e) If protein is present it adsorbs onto the indicator.

90 Which of the following possible urinary components would give a positive Ames Clinitest reaction and which would not?

(a) Lactose.
(b) Glucose.
(c) Ascorbic acid.
(d) Acetoacetic acid.
(e) Phenylpyruvic acid.

87 (a) **False** The M hydrochloric acid will inactivate the relevant enzymes.
(b) **False**
(c) **True** Hydrolysis of sucrose gives glucose and fructose.
(d) **True** The glucose oxidase test for glucose is dependent on the formation of gluconic acid and peroxide. Thus hydrogen peroxide will give a positive response.
(e) **False**

88 (a) **True** In (a), phenylpyruvic acid is being excreted.
(b) **False**
(c) **True** Phenistix contains Fe^{3+} ions and reacts with phenols. Aspirin (acetylsalicylic acid) is metabolized to form some salicylic acid

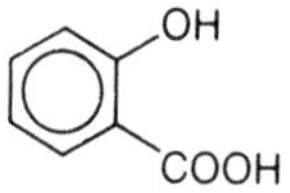

OH
COOH

which reacts with Fe^{3+} ions to give a purple colour.
(d) **False**
(e) **True** See (c) above.

89 (a) **False** There is no significant change in pH.
(b) **True** The pK_a of the bromophenol blue indicator is changed and a different colour is produced without a pH change.
(c) **True**
(d) **False**
(e) **True** This results in (b).

90 (a) **True** Reducing substances will give a positive Clinitest reaction.
(b) **True**
(c) **True**
(d) **False**
(e) **False**

91 Blood calcium
(a) Consists entirely of diffusible Ca^{2+} ions.
(b) Is involved in neuromuscular activity.
(c) Passes into the glomerular filtrate with no subsequent reabsorption from the kidney tubules.
(d) Has a concentration that can vary in the normal healthy adult by a factor of 2.
(e) Has a concentration that is controlled by hormonal action.

92 Bence Jones protein was detected in the urine of a patient. Which of the following statements would apply to this situation and which would not?
(a) Under electrophoresis, the protein ran as a γ-globulin.
(b) On heating, the protein coagulated, but appeared to dissolve again at about 65°C.
(c) The protein only coagulated when heated above 65°C.
(d) The patient is likely to have a condition called multiple myeloma.
(e) The patient must have kidney damage.

93 Within one community, a high proportion of new-born babies suffered from methaemoglobinaemia.

Which of the following statements might apply and be relevant to the situation and which would not?
(a) The community is exposed to industrial smoke.
(b) The local water supply is very rich in nitrites.
(c) The babies are suffering from the effects of mercury poisoning.
(d) Very high levels of nitrates contaminate the local water reservoir.
(e) The haemoglobin of the babies' blood is in a very reduced state.

91 (a) **False** A considerable proportion is protein-bound. In addition some may be present complexed in small molecular structures.
(b) **True**
(c) **False** Normally most of the calcium present in the glomerular filtrate is reabsorbed. It behaves as a threshold substance.
(d) **False** The normal range of serum calcium in adults is approximately 9 to 11 mg per 100 ml.
(e) **True** Parathormone and calcitonin are hormones controlling blood calcium levels.

92 (a) **True**
(b) **True**
(c) **False**
(d) **True**
(e) **False** Bence Jones protein is a very small protein and can filter through even if there is normal kidney function.

93 (a) **False**
(b) **True** High levels of nitrites can cause methaemoglobinaemia. Methaemoglobin can play no part in oxygen transport.
(c) **False**
(d) **True** Nitrates can be reduced by bacterial action to form nitrites, which could be responsible for the methaemoglobinaemia.
(e) **False** The haemoglobin is oxidized to methaemoglobin. Ferrous ion is oxidized to the ferric form.

94 Consider the oxygen uptake curve of haemoglobin.

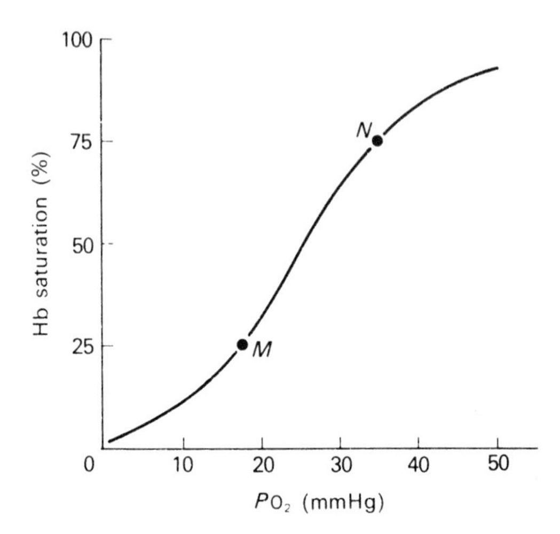

Which of the following can be correctly inferred from this curve and which cannot?

(a) At point *M* (25% saturation), one-quarter of all the haemoglobin molecules present are saturated with oxygen.
(b) At point *N* (75% saturation), each haemoglobin molecule has three oxygen molecules attached.
(c) In going from point *M* to point *N*, when the P_{O_2} approximately doubles, the haemoglobin is able to increase its oxygen uptake three-fold.
(d) At point *M* (25% saturation), one-quarter of all the available oxygen-binding sites are occupied.
(e) Haemoglobin, when saturated with oxygen, must be binding more than one molecule of oxygen per haemoglobin molecule.

95 Bile salts are:
(a) Synthesized in the liver from cholesterol.
(b) Another name for bile pigments.
(c) Necessary for normal fat absorption in the gut.
(d) Emulsifying agents.
(e) Entirely reabsorbed from the gut.

96 Which of the following are known to be involved in halting the flow of blood from damaged capillaries and which are not involved?
(a) Thrombin.
(b) Heparin.
(c) *p*-Aminobenzoic acid.
(d) Platelets.
(e) Calcium ions.

94 (a) **False** Although the oxygen uptake is cooperative and some molecules will be saturated, others will have 1, 2, or 3 oxygen molecules per molecular tetramer.
(b) **False** Cooperativity will imply that at point *N*, many haemoglobin molecules will be saturated, i.e., have 4 oxygen molecules in combination.
(c) **True**
(d) **True** However, these occupied sites are certainly not distributed uniformly throughout the molecules.
(e) **True** When saturated, 4 oxygen molecules will be bound per haemoglobin molecule.

95 (a) **True** Oxidation reactions and conjugation with glycine or taurine convert cholesterol into bile salts.
(b) **False**
(c) **True** The emulsifying action of bile salts is needed for fat absorption.
(d) **True**
(e) **False** Although reabsorption occurs, it is not complete.

96 (a) **True** Thrombin is a proteolytic enzyme that activates fibrinogen by removing peptide(s), thus enabling polymerization to occur to form fibrin.
(b) **False** Heparin is an anticoagulant present in mast cells located in connective tissue surrounding the walls of blood vessels.
(c) **False**
(d) **True** Platelets aggregate at a site of injury and release 5-hydroxytryptamine, which causes vasoconstriction. The aggregation forms a platelet plug.
(e) **True** Ca^{2+} ions are involved in several of the clotting stages.

97 The following materials are recognized human blood coagulation factors:
(a) Fibrinogen.
(b) Glutathione.
(c) Pepsinogen.
(d) Antihaemophilic factor A.
(e) Coenzyme Q.

98 The following comprise components that are present in the structures of some or all human connective tissue proteoglycans (mucoproteins).
(a) Acetylcholine.
(b) D-Glucuronic acid.
(c) Cholesterol.
(d) Serine.
(e) *N*-Acetyl-D-galactosamine.

99 Collagen
(a) Is a globular protein.
(b) Contains hydroxylysine.
(c) Is a polypeptide with glycine as every alternate amino acid.
(d) Contains hydroxyproline.
(e) Is present in bone.

100 Hydroxyapatite in dentine
(a) Contains carbonate.
(b) Contains citrate.
(c) Contains pyrophosphate.
(d) Never contains crystal defects.
(e) Is present in microscopically visible crystals.

101 Calcitonin
(a) Is a phospholipid.
(b) Is a polypeptide.
(c) Acts to reduce a rising plasma calcium concentration.
(d) Inhibits the action of osteoclasts.
(e) Is a product of the thyroid gland.

102 In the human body, iodine is
(a) Taken up by the adrenals.
(b) Taken up by the thyroid.
(c) Present in the amino acid thyroxine.
(d) Present in thyroglobulin.
(e) Converted by the liver into iodoacetamide.

97 (a) **True** Factor I of the blood clotting factors, it is a high molecular weight protein and is the precursor of fibrin.
(b) **False**
(c) **False**
(d) **True** Factor VIII of the blood clotting factors. Deficiency is the cause of classical haemophilia (haemophilia A).
(e) **False**

98 (a) **False**
(b) **True** The carbohydrate component of proteoglycans consists of chains of repeating disaccharide units. In several cases, β-D-glucuronic acid is one of the two monosaccharides making up each disaccharide unit.
(c) **False**
(d) **True** In general, polysaccharide chains will be covalently attached to a polypeptide chain by means of the HO-groups of serine or threonine.
(e) **True** *N*-Acetyl-D-galactosamine, or its 4- or 6-sulphate, form the second of the two monosaccharide units (in each disaccharide unit) in many of these mucopolysaccharides.

99 (a) **False** It forms fibrils.
(b) **True** This is a characteristic component of collagen.
(c) **False** Every third amino acid in a collagen chain is glycine.
(d) **True** Hydroxyproline is also a characteristic component of collagen.
(e) **True**

100 (a) **True**
(b) **True**
(c) **False**
(d) **False** Crystal defects are gaps in the hydroxyapatite crystal structure which would each contain the appropriate ion in the perfect crystal. Fewer defects are present in the fluoro-apatite crystal structure.
(e) **False** Hydroxyapatite is present in a submicrocrystalline form and in an amorphous form.

101 (a) **False**
(b) **True** It is a peptide of molecular weight *ca.* 3600 with 32 amino acids.
(c) **True** Among other actions it increases calcium excretion and inhibits synthesis of 1,25-dihydrocholecalciferol.
(d) **True**
(e) **True** It is believed to originate in the thyroid gland C cells.

102 (a) **False**
(b) **True**
(c) **True** Thyroxine is an iodine-containing amino acid derived from tyrosine.
(d) **True** This is the protein that contains iodinated amino acid residues.
(e) **False**

103 Which of the following statements apply to iron in the human body and which do not?
(a) It is readily excreted in the urine.
(b) It is present in both ferrous and ferric states.
(c) it is stored as ferritin and haemosiderin.
(d) It is a component present in the enzyme catalase.
(e) It is transported as ceruloplasmin.

104 Cytochrome P450
(a) Is an organel approximately 450 nm in greatest dimension.
(b) Is a coloured cell.
(c) Contains iron.
(d) Is an enzyme involved in certain oxidations.
(e) Is present in the microsomal fraction of liver cells.

105 Mixed function oxidases are important for the metabolism of many foreign compounds.
(a) They are enzymes found mainly in the kidney.
(b) They can be isolated from liver microsomal fractions.
(c) They are always mitochondrial enzymes.
(d) They always require ATP as a cofactor.
(e) They only carry out reactions involving aromatic ring hydroxylation.

106 When a drug or other foreign substance enters the body, it is metabolized and often conjugated with glucuronic acid, glycine or other substance before excretion. Consider which of the following statements would apply to this situation and which would not.
(a) The initial substance is likely to be made more water-soluble.
(b) After metabolism, the substance will be more highly ionized.
(c) The metabolites are more likely to penetrate lipid barriers.
(d) Any highly ionized metabolites will be reabsorbed in the kidney tubules.
(e) Metabolism will have no effect on the rate of excretion of the foreign substance.

103 (a) **False** Unless some iron-chelating drug has been administered, virtually no iron is excreted in the urine.
(b) **True** Iron in haemoglobin and myoglobin is essentially in the ferrous state, whilst iron in transferrin, ferritin and haemosiderin complexes is essentially in the ferric state.
(c) **True**
(d) **True**
(e) **False** This protein transports copper in the plasma.

104 (a) **False**
(b) **False**
(c) **True**
(d) **True** It is concerned in hydroxylations of the type

$$RH + O_2 + NADPH + H^+ \longrightarrow ROH + NADP^+ + H_2O$$

(e) **True** It is often concerned with the hydroxylation of drugs and other foreign substances.

105 (a) **False** They are mainly located in liver.
(b) **True** They are present in the endoplasmic reticulum.
(c) **False**
(d) **False** They require oxygen and NADPH.
(e) **False** They are involved with aromatic ring hydroxylations, but also in other oxidative processes such as dealkylation.

106 (a) **True** The general trend in such metabolism results in more water-soluble metabolites.
(b) **True** The ionized form is water-soluble, rather than lipid-soluble.
(c) **False** See (b) above.
(d) **False** Ionization and water solubility go together.
(e) **False**

107 Consider the antituberculosis drug, *p*-aminosalicyclic acid (PAS)

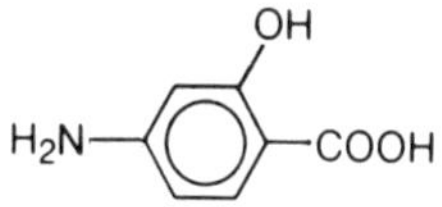

OH H_2N— —COOH

When administered orally to man, it is absorbed rapidly and undergoes metabolism before excretion.

Which of the following metabolic processes are likely to occur and which are not?

(a) Acetylation of the $-NH_2$ group.
(b) Conjugation with glycine.
(c) Oxidation of the $-NH_2$ group.
(d) Glucuronide formation.
(e) Reduction of double bonds.

108 Which of the following statements apply to the carcinogenic polycyclic hydrocarbons and which do not?

(a) They are never metabolized in the body.
(b) They require metabolic activation in the body before they exert their carcinogenic effects.
(c) The active carcinogen is considered to be an epoxide.
(d) They all form glycine conjugates before excretion.
(e) They all form nitrosamines in the body.

109 Which of the following statements are correct for drugs taken by mouth and which are not?

(a) Weakly acidic drugs may be largely absorbed from the stomach.
(b) All lipophilic drugs are quickly and readily excreted in the urine.
(c) Low molecular weight drugs cannot pass through the placental barrier.
(d) Weakly basic drugs are absorbed mainly from the intestine.
(e) Strongly basic drugs may undergo little absorption from the gut.

110 The following statements can be correctly applied to the action of peptide hormones:

(a) They can act by activating or inhibiting enzyme systems within the cell, without themselves penetrating the cell.
(b) They enter the cell linked to cyclic AMP and then influence enzyme systems.
(c) Each hormone has an action that is identical on all cells.
(d) The site of action of the hormone might be blocked by a competitive molecule.
(e) They are all large molecules with molecular weights above 10 000.

107 (a) **True** This is very common with $-NH_2$ groups.
(b) **True** This occurs through the $-COOH$ group.
(c) **False**
(d) **True** This can occur at the $-OH$, $-COOH$, or $-NH_2$ groups.
(e) **False**

108 (a) **False**
(b) **True** They are usually hydroxylated and form epoxides as intermediates.
(c) **True** These epoxides are reactive with DNA.
(d) **False** Only the $-COOH$ group forms glycine conjugates. This is not present with polycyclic hydrocarbons.
(e) **False**

109 (a) **True** In the acidic pH of the stomach, they are largely unionized and thus relatively lipid-soluble.
(b) **False**
(c) **False** Whether small molecular weight compounds are permeable or not depends largely on their lipid solubility.
(d) **True** In the alkaline pH of the intestine they are largely in the non-ionized lipid-soluble form.
(e) **True** Their lipid solubility may be small at any pH encountered in the gut.

110 (a) **True** ACTH molecules, for example, attach to receptor sites on adipose tissue cells and stimulate adenyl cyclase to convert ATP to cyclic AMP within the cells. The cyclic AMP can then activate the adipose tissue lipase system.
(b) **False**
(c) **False** Attachment to cell receptor sites is specific. Cells without appropriate receptor sites are unaffected by the hormone.
(d) **True** Thus the activity of the ACTH receptor sites on adipose tissue cells can be blocked by insulin.
(e) **False** Molecular weights of glucagon and ACTH for example are less than 5000.

111 The following statements refer to the subject of compartmentation within and without mitochondria.

(a) NAD^+ and NADH can readily cross mitochondrial membranes, whilst $NADP^+$ and NADPH cannot cross them.
(b) Acetyl-CoA is readily diffusible through mitochondrial membranes.
(c) α-Glycerophosphate and malate can transfer reducing power in the form of combined hydrogen through mitochondrial membranes.
(d) Citrate, being an ion of a tricarboxylic acid, cannot pass through mitochondrial membranes.
(e) Long-chain (fatty) acyl carnitine esters can be transferred across the inner mitochondrial membrane to the inner matrix compartment.

112 Which of the following statements concerning hormone action are true and which false?

(a) Steroid hormones cannot enter cells.
(b) Prostaglandins possess potent hormonal activity.
(c) Steroid hormones exert their primary effect on gene expression and, consequently, their activity depends on the synthesis of proteins.
(d) Inhibitors of phosphodiesterase, such as caffeine, can act synergistically with those hormones that act by stimulating the synthesis of cyclic AMP.
(e) In skeletal muscle, cyclic AMP activates a protein kinase that phosphorylates enzymes and, in consequence, glycogen phosphorylase is activated and glycogen synthetase is inactivated.

111 (a) **False** The mitochondrial membranes are virtually impermeable for these 4 molecules.

(b) **False**

(c) **True** These diffusible molecules link the oxidation–reduction status inside and outside mitochondria, e.g.,

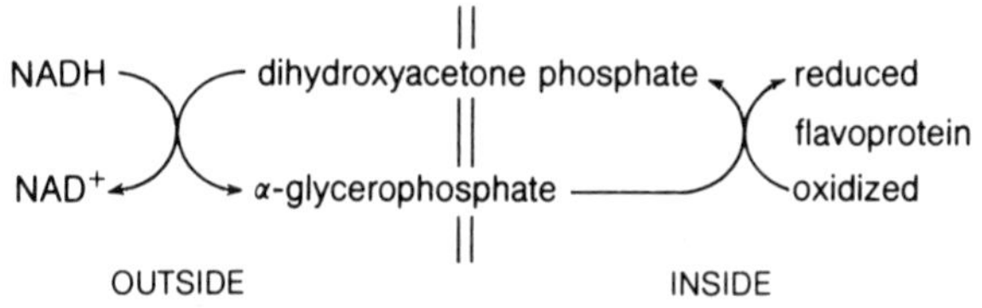

NADH → dihydroxyacetone phosphate → reduced flavoprotein

NAD^+ ← α-glycerophosphate → oxidized

OUTSIDE INSIDE

(d) **False** It readily crosses the membranes and effectively transfers acetyl groups.

(e) **True** Carnitine acts as the carrier for the transport of acyl-CoA derivatives of fatty acids into the mitochondria for oxidation.

112 (a) **False** Steroid hormones act at cell nuclei and require a mechanism for transportation through the cytosol and into the nucleus.

(b) **False** Prostaglandins modulate hormone action, but are not themselves hormones.

(c) **True** One result of this mechanism is that the activity of steroid hormones is comparatively slow to become apparent. Their full effect may take hours to develop.

(d) **True** This is because phosphodiesterase activity is a mechanism for hydrolysing cyclic AMP.

(e) **True** The dual enzyme effect prevents 'futile' synthesis occurring.

76

First published 1980
by Edward Arnold (Publishers) Ltd
41 Bedford Square, London WC1B 3DQ

Reprinted 1982

ISBN 0 7131 4363 0

Set by Preface Limited, Salisbury, Wilts
Printed in Great Britain by
Spottiswoode Ballantyne Ltd.,
Colchester and London